DEVELOPMENTAL AND CELL BIOLOGY SERIES

EDITORS

D. R. NEWTH AND J. G. TORREY

GENETICS OF HIGHER PLANTS
Applications of cell culture

GENETICS OF HIGHER PLANTS

Applications of cell culture

R. S. CHALEFF
Experimental Station
E. I. du Pont de Nemours & Co., Inc.
Wilmington, Delaware

FOREWORD BY
JOHN G. TORREY
Professor of Botany, Harvard University

CAMBRIDGE UNIVERSITY PRESS
CAMBRIDGE
LONDON   NEW YORK   NEW ROCHELLE
MELBOURNE   SYDNEY

Published by the Press Syndicate of the University of Cambridge
The Pitt Building, Trumpington Street, Cambridge CB2 1RP
32 East 57th Street, New York, NY 10022, USA
296 Beaconsfield Parade, Middle Park, Melbourne 3206, Australia

First published 1981

Printed in the United States of America
Typeset by General Graphic Services, Inc., York, Pa.
Printed and bound by Vail-Ballou Press, Inc., Binghamton, NY

Library of Congress Cataloging in Publication Data
Chaleff, R. S.   1947–
Genetics of higher plants.
(Developmental and cell biology monographs; 9)
Bibliography: p.
Includes index.
1. Plant genetics   2. Plant cell culture.
I. Title.
QH433.C48   582′016   80-18527
ISBN 0 521 22731 3

to Deborah

Contents

Contents

Foreword

In 1969 a small international group of plant scientists gathered in Bellagio, Italy, under the auspices of the Rockefeller Foundation; here were brought together for the first time all the essential elements of the field of somatic cell genetics of plants. From several laboratories in North America and Europe came the experience of plant tissue culture technology, from France the methods of meristem culture and vegetative propagation, from England and Sweden the techniques of cell suspension culture and protoplast production, and from Japan information on the purified and highly efficient hydrolytic enzymes needed for cell wall dissolution. From outside the field of plant tissue culture altogether came the geneticists interested in plant production, whose recognition of the need and expressed demand for new methods to broaden genetic crosses effectively catalyzed the reaction.

The reader will not find in this book the historical elements of the work that led to the establishment of the field of plant somatic cell genetics nor even, except in a cursory way, a discussion of the present technology; these historical and technological facts are largely assumed as familiar background by the author and can be found by the reader in the other volumes, such as the extensive review edited by H. E. Street, now published in an expanded second edition. Rather, the author has devoted himself to presenting a concise, up-to-the-minute account of the present state of the art of plant somatic cell genetics, the accomplishments to date, the major problems still to be overcome for further success to be achieved, and the potentiality of the methods for progress in the future. Developments have been rapid in the field of plant somatic cell genetics; much of the published work reviewed dates from 1970 onward, encompassing just a decade of research.

As befits an advanced treatise on a newly developing field, the author has chosen to use the language and terminology of the specialist. The book will be readily understood by the geneticist, the molecular biologist, and the plant biochemist. Yet its practical message is ultimately directed toward the plant breeder, the plant pathologist, and the agronomist. One finds in chapter after chapter clear exposition of complex information, careful weighing of sometimes controversial experiments or reports, and clear summaries of the lessons learned.

Full attention is given not only to the results of the experiments themselves but also to the practical uses and further implications of these experiments in such subjects as plant resistance to disease, plant tolerance to environmental stress, and improved nutritional values of plant food products.

Plant tissue culturists will feel that research of the sort summarized here is part of the long-term return from the basic research begun in the 1930s. There remain, of course, many blocks to accelerated progress in the field. Solution of some of these problems will come from further basic research in subject areas much less popular than this one. The author points to a few of these problems. Thus, for example, he says that "controllable regeneration of plants from single protoplasts of crop species" is one of the single greatest deterrents to progress. Success in overcoming this stumbling block will depend on a better understanding of the physiology and biochemistry of cytodifferentiation and morphogenesis. Another problem area concerns the genetic stability of cells in suspension culture and cells grown as callus tissue. Currently this difficulty is circumvented only by meristem culture, which for somatic cell genetic research is not an answer at all, even though "mericloning" has become an effective and useful tool for plant propagation, especially of horticultural species.

Taken altogether, Chaleff's book summarizes areas of tremendous progress, describing accomplishments not dreamed of in the early 1930s and outlining prospects for the future only now beginning to be visualized clearly. One can hope that this book will help to catalyze renewed efforts toward the resolution of unsolved problems and provide impetus for applications of these and further developments toward improved plant productivity.

Petersham, Massachusetts   John G. Torrey

Preface

Plant cell and tissue culture is not a new science. P. R. White first grew tomato roots in a liquid nutrient medium in 1934, and shortly thereafter R. J. Gautheret established callus cultures of goat willow and carrot. Many recent volumes have summarized the advances realized in this field, but these reviews have emphasized the application of cell and tissue culture techniques to physiological studies, vegetative propagation, and the generation of virus-free plant material. None has rigorously examined the potentiality of cell culture as a method for accomplishing genetic modifications of higher plants. Experimentation with plants contributed greatly to the early development of the science of genetics. But the biological complexity of these multicellular forms and their long generation times encouraged genetics researchers to shift to the study of the more experimentally approachable microorganisms. Yet the task of elucidating the organization of these higher forms must be confronted. In the final analysis, all life is dependent on photosynthesis, and not until an understanding of the molecular processes of plants is attained can these forms be successfully manipulated to improve the quality of that life.

The ability to culture cells of higher plants on chemically defined media and to regenerate plants from these cells offers new possibilities for genetic experimentation. Direct selection for defined mutant types, an experimental advantage previously restricted to microbes, can now be accomplished with cultured plant cells. This development places us on the threshold of what must prove to be a new realm of scientific discovery.

It is impossible for one person writing in the infancy of this exciting new field to predict experimental directions or capabilities. This modest volume is intended to summarize and evaluate the achievements to date, to provide a perspective of that work, and to encourage further efforts. It is written for those with some background in the biological sciences who entertain an interest in plant cell genetics, for students contemplating where to invest their talents, and for those actively engaged in the field.

Only scientific research that is no longer progressing and has been fossilized into dogma is free of debate and controversy. The opinions presented here are intended not to be judgmental but rather to provide a stimulus to further discussion and critical research.

This book can present only a glimpse taken at one moment in time of a dynamic and ever-developing area of research. As such, it is inevitable that it be obsolete in some respects by the time of publication because of advances made during its preparation. I have tried to include information from relevant publications that appeared in print before late 1979.

I am most grateful to Drs. Deborah Chaleff, Gerald Fink, Ralph Keil, Kanak Samaddar, and Adrian Srb, whose valuable criticisms and suggest-sions have transformed my inchoate manuscript into this final text. I also would like to thank those of my colleagues who generously made their unpublished experimental results available.

June, 1980   R. S. C.
Wilmington, Delaware

And he gave it for his opinion, that whoever could make two ears of corn or two blades of grass to grow upon a spot of ground where only one grew before, would deserve better of mankind, and do more essential service to his country . . .

Jonathan Swift
Gulliver's Travels

1

Introduction

The modern science of genetics began with the study of plants. Mendel's formulation of the fundamental laws of genetics was based on his work with peas *(Pisum sativum)*. And, although Mendel's discoveries at first were ignored, they were brought to light in the year 1900 by Correns, DeVries, and Tschermak and verified through experiments with a diversity of plant genera. Plants remained a major subject of genetic investigation until the 1940s when methods were established for isolating mutants and for effecting controlled matings and genetic recombination in microbes. Several characteristics of microbial systems afford irresistible advantages that quickly led to their displacing higher plants as the favorite child of the geneticist. Mutant selection and characterization are greatly facilitated by the ability to grow microbial cells in a controlled environment. Thus, microbial cells can be plated on a simple chemically defined medium to which nutritional supplements or selective agents can be added. The introduction of such media made possible the isolation of auxotrophic mutants that are unable to synthesize certain metabolites required for growth. By means of modifications in other parameters of the in vitro habitat, such as temperature, additional types of conditional lethal mutants can be recovered that grow in one environment but not in another. In addition to furnishing markers for use in crosses and complementation tests, such mutants are valuable tools for dissection of biosynthetic and developmental sequences and of the regulation of gene expression.

The haploidy typical of the fungal and bacterial genomes also represents a desirable feature to the geneticist. Because a microbial cell contains only a single copy of each gene, recessive mutations, the most frequent kind, are not concealed by a normal allele as they often are in a diploid organism. Instead, new alleles are expressed directly by the vegetative descendants of the mutated cell, and sexual crosses are not required for their detection. Because mutations are rare events, the search for variability is a pursuit concerned with numbers. Therefore, another advantage of microbes is that large populations are readily produced vegetatively, and hence large numbers of genomes can be produced in a small volume. Furthermore, such a population may be

1

essentially homogeneous in that all of its members are in the same developmental or physiological state. Because of this feature of microbial cultures, growth conditions can be defined to elicit expression of that function in which a mutation is desired, and novel phenotypes that arise can fairly safely be assumed to have a genetic basis. In addition, although one is able to work with enormous populations of microorganisms, their growth as discrete units facilitates other important experimental manipulations. For example, the capacity of isolated microbial cells to multiply permits cloning to be accomplished easily. As single cells also have only small endogenous reserves of nutrients that are rapidly depleted when the ability to manufacture a particular metabolite is lost, the lag in expression of such a deficiency is minimal. The short generation times of microbial cells permit the results of genetic experiments and growth tests to be obtained in 1 or 2 days and masses of cells to be produced for biochemical analyses in but a short time. Here in the microbe we have the ideal organism for genetic experiments. It is no wonder that research with bacteria and fungi has contributed so much to our knowledge of genetic function and organization and has advanced so rapidly to such a high degree of refinement and sophistication.

Plants could not compete as experimental genetic systems with the array of advantages possessed by microorganisms. In contrast to a microbial culture (10^8 cells per milliliter), a field of plants offers only a relatively small number of individuals (approximately 10^4 plants per acre) to be screened for variation. Moreover, these plants are complex associations of highly differentiated cells that are growing on a substrate that is both undefined and uncontrolled. As a result, few methods are available for selecting directly for mutations in specified loci, and one is largely restricted to identifying mutants solely on the basis of altered appearance (e.g., color or morphology). Once a mutation is obtained, genetic analysis demands that it be present in the germ line and that one endure the entirety of a very lengthy life cycle. And so an eclipse came over the plant kingdom.

Potentialities of cultured plant cells as a genetic system

Now while geneticists were indulging their predilection for microbial flora, many physiologists were developing techniques for culturing cells and tissues of plants. Through their efforts, conditions were elaborated for propagating plant cells in vitro as a mass of undifferentiated tissue (callus) on defined media. Callus can be formed from tissue explants or by culturing populations of single cells or protoplasts. By means of transfer of callus to a medium of appropriate hormone composition, adult plants can be regenerated. These operations, depicted in Figure 1.1, became

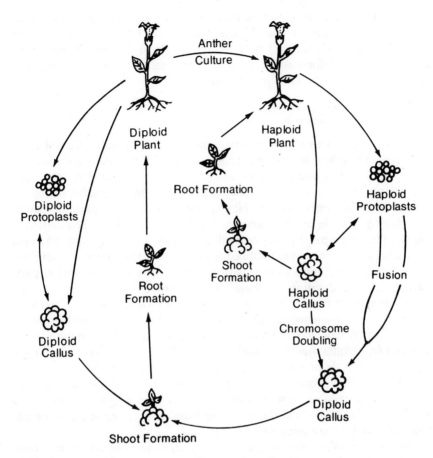

FIGURE 1.1. Summary of experimental manipulations possible with *Nicotiana*, *Datura*, and *Petunia*.

practicable with many plant species, and with a small number of species it was even possible to promote the formation of haploid plants from pollen by placing immature anthers in culture. Many researchers perceived that the introduction of these new methods conferred on plants attributes that had made microorganisms so attractive to geneticists. With the availability of large homogeneous populations of haploid plant cells, it seemed a simple matter to select defined mutants by incorporating toxic or growth-inhibiting compounds in the medium. Genetic analyses then could be performed by conventional methods with regenerated diploid plants. The ability to form plants from cultured cells further suggested the possibility of applying cell culture to introduce desirable genetic modifications of agronomically important traits of the whole plant that could not be accomplished readily or at all by traditional plant breeding methods. The new techniques of plant cell culture gave rise to prodigious expectations. Prometheus went forth to bring molecular genetics from the Olympus of modern science to the kingdom of plants.

Difficulties to be mastered

The confidence with which genetic experimentation with cultured plant cells was at first approached was engendered by the dramatic successes of microbial genetics and the naive reductionist conviction that any single cell would behave as a microorganism. But, in contradistinction to microbes, cells composing a multicellular organism evolve and function as a system of cells in intimate association with one another. The cells of a metaphyte specialize by differentiation, so that each cell type performs only a small number of the multifarious functions involved in forming and maintaining the complex organization and structure of which it is part. Moreover, coordination of these diverse processes requires that the component cells be able to communicate. Doubtless plant cells perform many of the same elemental activities and therefore have many features in common with unicellular organisms. However, as the metaphyte is the product of cellular functions and relationships that are unknown to the microbe, the cells of the metaphyte correspondingly must possess some properties and capabilities very different from those of the microbe. It is because of several of these peculiar characteristics of plant cells and our poor understanding of them that our aspirations for genetic experimentation thus far have not been well fulfilled.

Unlike microbial cells, which can multiply when placed in an infinite volume of medium, cultured plant cells require a minimum inoculum density to achieve self-sustaining growth and division. Although the initial population density that will support cell proliferation varies according to the origin and history of the cell line and the composition of the medium, typically it is in the range of 10^4 to 10^5 cells per milliliter. However, a single isolated plant cell will divide when placed on a filter paper resting atop a nurse callus (Muir, Hildebrandt, and Riker, 1958). This same growth-promoting diffusion effect is observed when a Petri dish containing a low density of cells in an agar medium is inoculated with a piece of callus tissue. Division begins first in the single cells nearest the callus and then progressively advances to more distant cells (Street, 1968, 1977). The ability of the nurse callus to supply diffusible substances that can induce and support division of the single cell suggests that the high inoculum density required for growth of a cell population is caused by permeability of the cell membrane. When a sparse cell population is cultured in a simple medium, the continuous efflux of essential metabolites will reduce their endogenous concentrations below the threshold levels necessary for cell growth. But this efflux will abate or cease when adequate amounts of these metabolites are established in the medium. In a dense population of leaky cells, an extracellular concentration of metabolites sufficient to retard further diffusion may be reached before the intracellular pools have been too depleted to sustain growth. It is expected, therefore, that addition of these required metabolites to the medium will reduce the initial cell density required for growth.

One method of enriching a medium with these compounds is first to use it to support growth of a dense cell suspension. The "conditioned" medium then is separated from the cells by filtration. The effective initial cell density for a conditioned medium is lower than for the pristine medium (Stuart and Street, 1969, 1971; Street, 1977). Growth-promoting substances also are released into the medium by dead cells. Nondividing X-irradiated protoplasts have served as feeder cells in supporting division of low densities of viable tobacco protoplasts (Raveh, Huberman, and Galun, 1973). The required minimum inoculum density can also be reduced by directly supplementing the medium with defined compounds. The addition of amino acids, nucleic acid bases, several sugars, sugar alcohols, and organic acids to a minimal medium of mineral salts, sucrose, glucose, vitamins, and 2,4-dichlorophenoxyacetic acid (a synthetic auxin) successfully decreased the minimum effective density of *Vicia hajastana* cells from 1200 cells per milliliter to 25 to 50 cells per milliliter. When casamino acids and coconut milk were included in this medium, growth was obtained from an initial population of 1 to 2 cells per milliliter. Except for their intolerance of free amino acids and nucleic acid bases, the response of *V. hajastana* protoplasts to these supplements was similar to that of cells. A single protoplast introduced into 4 ml of this enriched medium proved capable of forming a mass of approximately 10^7 cells within 30 to 40 days (Kao and Michayluk, 1975). So it appears that single plant cells can divide and grow in vitro, albeit in a complex and chemically undefined medium.

Another feature of plant cells that hinders genetic experimentation is their tendency to grow in culture as aggregates. A population of single cells is obtained only by filtration or conversion to protoplasts by treatment with enzymes. However, conditions have not yet been defined to obtain division of single cells or protoplasts of many species, most notoriously cereals (King, Potrykus, and Thomas, 1978). Nutrients or chemical agents intended for mutant selection cannot penetrate uniformly to all cells composing the aggregate. Consequently, cells in the interior of the aggregate will not be exposed to as high a concentration of the selective agent as will the exterior cells. Some nonmutant cells, therefore, might escape growth inhibition because of their location in the aggregate rather than as a result of a genetic modification that confers resistance. In such conditions selection cannot be stringent, and an altered growth response must be confirmed by further testing.

The efflux of metabolites from the members of an aggregate or a dense cell population impedes the selection of nutritional mutants. Auxotrophs that find their required nutrient in the medium will grow normally. In addition, in experiments in which mutants are selected on the basis of their ability to overproduce certain metabolites, these metabolites may be excreted by the mutants, thereby permitting normal cells in their proximity to survive. Such cross feeding will lead to the formation of chimeral calluses composed of normal and mutant cells. For example, amino acid analogues or amino acids that are the end products of a branched pathway will prevent growth of normal cells by inhibiting the activity of enzymes involved in the synthesis of specific amino acids (see Chapter 4). Mutants possessing enzymes insensitive to this inhibition will grow, but so perhaps will normal cells bathing in the excess of the amino acid released by the mutant cells. Of course, unless a pure mutant cell line is obtained by cloning single cells (by use of either a nurse callus or conditioned medium), biochemical analyses of the callus tissue or of regenerated plants will have neither sense nor value. Regeneration of plants from chimeral callus could yield chimeral plants, or perhaps, if the mutation interferes with organogenesis, normal plants would be recovered preferentially. Cloning can also be accomplished by having mutant cells incorporated in the germ line of a regenerated plant and passed through a sexual cross. This procedure was employed to obtain pure lines of a glycerol-utilizing mutant of *Nicotiana tabacum*. The formation of both normal and mutant plants from the glycerol-utilizing callus culture suggested that it was chimeral (Chaleff and Parsons, 1978a). Since cultured plant cells excrete glucose-containing polysaccharides (Becker, Hui, and Albersheim, 1964; Aspinall, Molloy, and Craig, 1969; Olson et al., 1969), it is probable that the mutant cells were converting glycerol into polysaccharides and sugars that were released into medium and used as carbon sources by normal cells. The design of the selection scheme is critical in determining if chimeral cell lines will be recovered, and it should be possible in most cases to predict whether or not normal cells will survive selection by cross feeding.

The growth of plant cells as aggregates also interferes with attempts to quantitate frequencies of mutation and survival after mutagenesis. Even though most of the cells composing an aggregate may be unable to grow on a particular medium or following a given treatment, a colony could be formed by the remainder, giving the same result as if all the cells of the aggregate were capable of growth.

The inability to grow isolated single plant cells in a simple and completely defined medium makes difficult the manipulations that can be performed so easily with microbes. But an important advance has already been made toward removal of this experimental constraint. When tobacco protoplasts that had been cultured for 4 days at a high density were diluted in a simple medium containing a reduced auxin concentration, cells could grow at a concentration as low as 1 to 2 cells per milliliter (Caboche, 1980).

In the following chapters the reader will be familiarized with other trying characteristics of cultured plant cells. The generation time for plant cells in cultures is rather long (approximately 30 hours) in comparison with that for a microbe (20 minutes for *Escherichia coli* and 90 minutes for yeast). Hence the induction of callus, the formation of cell suspension cultures, the selection of variants, and the performance of growth tests are time-consuming operations even when considered separately, but patience becomes a most important desideratum when they must be conducted sequentially, as is often the case. At present, the only method of performing a genetic analysis is by conventional sexual crosses with regenerated plants, and the regeneration of plants from callus cultures takes several months in the limited number of species in which it can be accomplished. Moreover, for most species in which regeneration of plants from callus is feasible, the morphogenetic capacity of the callus declines rapidly with continued propagation (see Chapter 3). Thus it often is not possible to subject a variant cell line to sustained selection in order to confirm its phenotype or to purify it by eliminating nonvariant cells. This inability to control the differentiation of plants from cultured cells is perhaps the single most serious obstacle to genetic studies.

The geneticist entering the plant kingdom is also faced with inadequate knowledge of metabolic pathways and their regulation. Without such information, the tasks of devising selection methods and of analyzing mutants are much more difficult. Then there is the problem of polyploidy. The polyploid nature of most plant species at best complicates and at worst makes impossible the production of monohaploid cells for isolating recessive mutants.

The difficulties described now to some degree frustrate and hinder genetic research with cultured plant cells. But they will be confronted and eventually solved. They were, after all, to be expected. Whereas one studies microbial systems because of features that suit them ideally to genetic experimentation, one elects to work with plants because of their overriding importance to mankind. But because plant cell genetics is inevitably examined in the light shed by the advances in microbial genetics, its present primitive state appears egregious. Instead, the limited progress in plant cell genetics should be viewed historically and understood as being due to the current stage of technical development of plant cell culture, which confines genetic studies with plant cells to a level comparable to that of microbial systems 20 to 30 years ago. Even advances in more tractable genetic systems depended on technical innovations. *Neurospora* genetics would have followed a different course if the isolation of mutants and the analysis of crosses had continued to be performed by establishing clones from ascospores in separate culture tubes. The ability to develop individual colonies from many ascospores or conidia on a single Petri dish resulted from the discovery by Tatum, Barratt, and Cutter (1949) that restricted colonial growth of mycelia is induced by sorbose. Similarly, for many years the technique of genetic transformation was confined to a few species of bacteria and even in these cases was unpredictable because of its dependence on the state of competency of recipient cells to accept donor DNA. However, the finding that calcium ions permeabilize the bacterial membrane to exogenous DNA facilitated the use of transformation for genetic mapping and extended the technique to other species such as *E. coli* (Mandel and Higa, 1970; Cosloy and Oishi, 1973). Genetic analysis of cultured animal cells, which relies on the resolution of somatic cell hybrids, would not have been feasible without the discovery that fusion of these cells is promoted by inactivated Sendai virus.

We must be confident that progress will be realized with plant cell culture methods that will provide a substratum on which advanced genetic systems can be built. Even if the difficulties are more formidable than those encountered in microbial systems, the motivation to overcome them is also greater, for what we have to gain from the development of plant cell genetics as a science is an enormously enhanced capability to manipulate the genetic composition of crop species. And surely no human being can experience a greater sense of satisfaction or fulfillment than that derived from seeing the fruits of his efforts feed his fellow man.

2
Genetics versus epigenetics

Plant development and epigenetics

One of the most intriguing enigmas facing modern science is the process by which a single-cell zygote gives rise to the population of diverse differentiated cell types that compose the multicellular eukaryotic organism. Experimental evidence from several sources supports the belief that differentiated cells possess the entire complement of genetic information present in the zygote. Cytological observations have shown that the process of nuclear division ensures that daughter cells normally receive identical sets of chromosomes. Even the divergent types of differentiated cells produced by successive cell divisions are seen to contain the same chromosome complement as the zygote. The classic experiments of Gurdon (1960) demonstrated that nuclei from differentiated tissues of the frog, when transplanted into an enucleated egg, are capable of promoting the development of an adult animal. This same principle operates in the plant kingdom. Working independently and at very nearly the same time, Steward, Mapes, and Mears (1958) and Reinert (1958, 1959) successfully produced mature plants from cell cultures derived from differentiated carrot tissues. Subsequently it was shown that the formation of a complete and fertile plant could proceed from a single isolated somatic cell (Vasil and Hildebrandt, 1965).

Cells that are capable of developing into a complete adult individual are said to be totipotent. The number of examples of totipotent cells has increased steadily since these initial discoveries, and it is tempting to generalize to the totipotency of all cells. Yet, at this time, all that can be said is that the totipotency of some specialized somatic cells of animals and plants has been verified. Nevertheless, the experimental demonstration of the totipotency of differentiated cells proves that, in some cases at least, the processes of development and differentiation do not involve loss or irreversible inactivation of genetic information from the cells but rather the regulation of expression of this information. A more trenchant formulation of this concept was provided by Wardlaw (1970) when he wrote that "all of the genes may function some of the time, and some of them all of the time, but not all of them all of the time."

9

The regulation of gene expression in plants has been demonstrated directly at the molecular level by Goldberg and associates (1978). In these experiments, messenger RNA (mRNA) that was being transcribed actively was isolated from polysomes of tobacco leaf tissue. An excess of this mRNA was hybridized with a population of radioactively labeled DNA molecules that represented all of the structural genes contained within the tobacco genome (single-copy DNA). Approximately 5 percent of the DNA sequences hybridized to the mRNA, showing that only this small fraction of the total genetic information of the tobacco genome is expressed in the young leaf. Although the leaf is a complex organ composed of several differentiated cell types, it can be assumed that these results primarily reflect transcriptional activity of the parenchymal cells, the predominant cell type in the leaf. Of course, some of the genes expressed in the leaf cells control basic metabolic functions common to all cell types, but others encode structural proteins and enzymes that are peculiar to the leaf. In other cell types, also, only 5 percent of the total genome may be expressed, and although it should be expected that some of these genes will be the same as those that are active in the leaf, others will be different and characteristic of that particular cell type. Thus at each stage of development or differentiation only a small fraction of the total genome may be expressed in any given cell type, but the spectrum of genes expressed is unique to that cell type.

The postulate that the processes of development and differentiation involve the sequential and differential expression of genes was first asserted by Haldane in 1932. The term "epigenesis" is generally used to describe this regulation of gene activity that occurs during the course of development (Waddington, 1940). However, since biological systems are not static, but at all times are responsive, dynamic, and developing, they will involve changes in gene expression other than those defining the normal course of development and differentiation. Nanney (1958) extended the definition of the term "epigenetic" to include all changes in the regulation of gene expression that may persist indefinitely throughout cellular divisions following removal of the inducing conditions. This characteristic stability of epigenetic changes distinguishes them from physiological changes that appear in response to a stimulus and disappear following cessation of that stimulus.

Possible mechanisms of epigenetic regulation

Epigenetic control of gene activity may be accomplished by selective regulation of several processes: transcription of RNA from DNA, processing of RNA and its transport into the cytoplasm, degradation of mRNA, and translation of mRNA into protein.

An example of transcriptional control in plants is found in the regulation of synthesis of the large subunit of ribulose bisphosphate carboxylase (RUBPCase) in chloroplasts of *Zea mays*. In maize (and other C4 plants) this polypeptide is found in chloroplasts of bundle sheath, but not in those of mesophyll cells (Huber, Hall, and Edwards, 1976). When the gene that encodes the large subunit of RUBPCase is isolated from plastid DNA and cloned in a bacterial plasmid (Coen et al., 1977), the purified nucleotide sequence can be used as a molecular probe in hybridization experiments to detect the presence of the RUBPCase gene in the genomes of chloroplasts of the two cell types and the presence of complementary RNA sequences within the chloroplasts. Although the DNA sequence coding for the large subunit of RUBPCase is present in the genomes of both mesophyll and bundle sheath chloroplasts, RNA complementary to this sequence has been detected only in bundle sheath cells (Link, Coen, and Bogorad, 1978). That a gene present in chloroplasts of mesophyll and bundle sheath cells is transcribed in one cell type and not in the other is convincing evidence for control of gene expression at the level of mRNA synthesis.

Regulation of processing and transport of mRNA was first suggested by the discovery in the nuclei of animal cells of a class of heterogeneous nuclear RNA molecules (HnRNA) that are larger and have greater sequence complexity than mRNA. Most of the HnRNA is degraded within the nucleus, and only a small fraction enters the cytoplasm. The similar base compositions of HnRNA and mRNA and the fact that polyadenylic acid sequences are attached to the 3' end in a large proportion of both HnRNA and mRNA molecules have led to acceptance of the hypothesis that HnRNA is a precursor of mRNA (as reviewed by Lewin, 1974). Recently, Goldberg and associates (1978) identified in nuclei of tobacco leaf cells a class of RNA molecules that are larger than the mRNA isolated from cytoplasmic polysomes. The numbers of unique nucleotide sequences in the nuclear and mRNA populations were compared by measuring the extent of hybridization with radioactively labeled single-copy DNA. The nuclear RNA population contains approximately 19 percent of the single-copy sequences present in the tobacco genome, whereas only 5 percent of the sequences are represented in the polysomal mRNA. The approximately fourfold greater sequence complexity of the tobacco nuclear RNA is in close agreement with the differences found between HnRNA and mRNA of animal cells (Hough et al., 1975; Bantle and Hahn, 1976; Levy, Johnson, and McCarthy, 1976). These results indicate that although a substantial proportion of the plant genome is transcribed (HnRNA), only small numbers of these sequences are transported to the cytoplasm (mRNA). Thus it appears that degradation of RNA within the nucleus and/or selective transport through the nuclear membrane are additional mechanisms by which gene activity may be regulated in higher plants.

Gene expression may also be regulated by controlling the time at which an mRNA molecule is translated into protein once it has entered the cytoplasm. Translation may be prevented or delayed by sequestering the mRNA in an inactive but stable form or by reducing the affinity between the mRNA and the translation apparatus. That at least one of these mechanisms operates in higher plants is suggested by studies of the effects of inhibitors on RNA and protein synthesis in germinating seeds. Application of cycloheximide to cotton seeds during the early stages of germination arrests both protein synthesis and embryo development. However, treatment with actinomycin D inhibits RNA synthesis without affecting either protein synthesis or the growth rate of the embryos (Waters and Dure, 1966). Hence, protein synthesis in the early stages of cotton seed germination employs RNA templates and ribosomes that were formed previously and stored in the seed.

From this brief discussion it is evident that the final phenotype of a cell is realized through the sequential and selective expression of genes throughout development and that the differentiated state is maintained by the continued activity of but a small fraction of a genome that remains structurally unaltered. In higher plant cells epigenetic control of gene function may be accomplished at any of the several steps involved in directing the synthesis of a protein molecule from information encoded by the DNA.

Changes in epigenetic regulation of gene activity must be regarded in contrast to and as distinct from changes caused by genetic events. Genetic events that can produce an altered phenotype include the following: substitution, insertion, and deletion of nucleotides; recombination, inversion, and translocation of chromosome segments; and change in chromosome number.

Two corollaries of epigenetic regulation will now be considered, with several examples that illustrate their relevance to genetic studies with cultured plant cells.

First corollary: expression of epigenetic changes by cultured cells

The expression of a new phenotype by cultured cells may result from an epigenetic modification rather than a genetic modification. The variant phenotype may reflect the expression of information that is present in the plant genome but that normally is not expressed in cultured cells. In such a case the rate of synthesis of a gene product is altered, but the nucleotide sequence of that gene is not.

The earliest observations of epigenetic control in cultured plant tissues were of modifications of the hormonal requirements of cell proliferation.

Gautheret (1946) discovered that *Scorzonera* tissue, which normally requires an exogenous supply of auxin, occasionally, following several passages in culture, acquired the ability to grow without an auxin supplement (called auxin autotrophy for lack of a better term). These auxin autotrophic cell cultures no longer responded to the addition of auxin to the medium, and they differed morphologically from hormone-dependent tissue. Likewise, carrot cultures that did not require exogenous auxin for growth, but that were responsive to auxin supplementation, proved capable of developing an insensitivity to the presence of this hormone in the medium. This phenomenon subsequently was observed with cultured tissues of many species (as reviewed by Gautheret, 1955). Gautheret (1946) referred to such reductions in the sensitivity of cell cultures to auxin as *accoutumance à l'hétéro-auxin*. Authors writing in English have translated this term as "habituation" (White, 1954), a more general designation that also can be applied to the loss of a cytokinin requirement (Binns and Meins, 1973). The habituated state is extremely stable in that succeeding generations of cells exhibit the same phenotype, and hormone dependence is rarely regained in culture (Gautheret, 1946). Since habituated cells contain elevated endogenous concentrations of the hormone for which they are habituated (Fox, 1963; Dyson and Hall, 1972), it was presumed that such changes result from an altered rate of hormone synthesis or degradation. The question that then had to be answered was whether a genetic or epigenetic event is responsible for the transition to hormone autotrophy.

The first evidence suggesting that habituation is an epigenetic phenomenon came from studies on the reversibility of the hormone-autotrophic phenotype. The differentiation of plants will obliterate the capacity for growth in the absence of exogenous hormones. Thus, new callus cultures initiated from plants that had been regenerated from an auxin-habituated tobacco culture (Sacristán and Melchers, 1969) and from an auxin- and kinetin-autotrophic cell line of *Crepis capillaris* (Sacristán and Wendt-Gallitelli, 1971) were found to require both auxin and kinetin for growth. However, as cell cultures are genetically heterogeneous and certain genotypes selectively regenerate (Sacristán and Melchers, 1969), it was not proved that regeneration induced loss of the habituated phenotype. The possibility remained that habituation resulted from a mutation and that the plants examined had originated from normal hormone-dependent cells contained within the population or from cells in which the mutant allele had reverted.

The only means by which to resolve the question of habituation was by examining large numbers of autotrophic clones derived by isolation of single cells from a habituated culture. Although reversion still might occur during multiplication of the cloned cell lines, because such events are rare not all clones would behave alike if this were the case. However, if habituation has an epigenetic basis rather than a genetic basis, the same transformation might be observed for all clones. Binns and Meins (1973) performed these experiments with a cytokinin-habituated cell line of *Nicotiana tabacum*. Plants were regenerated from 19 cytokinin-autotrophic clones of single-cell origin, and callus derived from pith explants of all plants required cytokinin for growth. It was concluded from these results that habituation involves a change in the expression of genes that control endogenous hormone concentrations. This altered pattern of gene expression persists qualitatively, albeit not necessarily quantitatively, in daughter cells produced by mitotic divisions (Meins and Binns, 1977) and is reversed by the processes of differentiation and/or dedifferentiation.

The enchanced tolerance to chilling exhibited by selected cell lines of *N. sylvestris* may provide another example of an epigenetic alteration. These cell lines were selected for growth at 25°C after incubation for 3 weeks at 0°C or −3°C (Dix and Street, 1976). Following serial subculture at 25°C, three selected cell lines proved more resistant than normal control cultures to the effects of cold treatment on cell growth (Dix, 1977). However, callus derived from plants that had been regenerated from these three selected cell lines showed normal sensitivity to chilling.

Resistance to the herbicide picloram by one of seven selected cell lines of *N. tabacum* also may have resulted from an epigenetic event. Although this cell line continued to exhibit resistance to picloram following long periods of growth in its absence, only sensitive cultures were obtained from 10 regenerated plants. A genetic basis for resistance could be assigned to four other cell lines, since plants regenerated from these cell lines gave rise to resistant callus cultures, and in crosses with each of these regenerated plants resistance was transmitted as an allele of a single nuclear gene (Chaleff and Parsons, 1978b).

From the experiments with habituated cultures it is obvious that stability of a variant phenotype in culture in the absence of selection is not an adequate criterion by which to distinguish epigenetic and genetic events. The cases of chilling resistance and of the one picloram-resistant cell line that have been described also may have resulted from epigenetic traits that are expressed immutably by successive cell generations in culture. But until a study is made of clones derived from these cell lines, it cannot be known that loss of the altered phenotype simply did not reflect the preferential regeneration of plants from normal cells contained within the population. These examples seem to argue that if cells are cloned as they were in the experiments with habituated cultures, persistence of an altered phenotype in callus cultures derived from regenerated plants could provide sufficient proof of a true genetic change. However, these examples are few in number, and it must be remembered that the metabolic

given phenotypic change. The rate of appearance of that new
type then would be 10^{-4}, which might persuade some that the
rmation is epigenetic. Thus, in the absence of any information
the number of loci that can be involved, rates and frequencies are
ngless. Interpreting such data would be like trying to make sense of
ion without knowing the denominator. Second, several features of
owth of plant cells in culture make estimation of the rate or
ncy of phenotypic change problematic. These technical difficulties
ussed in Chapter 4. However, it should be mentioned en passant
eriments designed to determine the effect of a mutagen on the rate
earance of a given phenotype should include a control that
trates the potency of the mutagen. To my knowledge, this simple
has not been performed in any such experiment that has been
d. The effect of the mutagen on the mutation rate of a character-
terial or yeast gene would serve as an easy and direct test of its
ic activity. Otherwise, of course, one does not know whether the
of a treatment to effect an increase in rate results from the
ic nature of the change or from the inactivity of the mutagen.

ursory consideration of genetic and epigenetic events leads us
ly to the conclusion that at the present time, transmission
sexual crosses is the sole valid criterion by which they can be
shed. But the diversity of biological systems should have taught
e should not rush to embrace any absolute standards for their
n. The strength of formal genetic analysis is that it is based on
ergence of many different lines of experimentation: predicted
n patterns from crosses, mutation and reversion frequencies,
mutagens, recombination, and even DNA sequencing. One
ntal approach may be more convenient in a given system, but
ld be singled out as having more merit than another. Instead,
s must be derived from and must be consistent with evidence
y several types of experiments. We must also remember that as
are refined and the resolution of our analytical methods
new discoveries will be made that will change our concepts and
at definitions and standards be rewritten or abandoned. Per-
netic events that are transmitted through meiosis and genetic
at are not visible in subsequent generations may yet be
Nevertheless, for the moment, an operational definition would
the term "mutant" be reserved exclusively for those cases in
netic basis for an altered phenotype has been established
lly by studies of inheritance across sexual generations. If an
notype is not heritable (in which case it is epigenetic) or if a
lysis has not been performed, cell lines exhibiting that
hould be referred to as "variants."

function altered by habituation, maintenance of endogenous concentrations of hormones, is critical to differentiation and that differentiation
might proceed only if normal hormone levels are restored. It is possible
that other cases of epigenetic change might involve altered expression of
more general cellular functions that are also active in differentiated tissue.
Such changes, although epigenetic, might endure through a cycle of
differentiation and dedifferentiation. Therefore, the only acceptable
criterion that can be applied to distinguish genetic and epigenetic events is
the inheritance of a trait by progeny produced by sexual crosses.
Transmission through the gametes indicates that a trait is maintained
stably through meiosis as well as through mitosis and is the best possible
evidence of a structural change in the DNA. In cases of extra-
chromosomal inheritance, the analysis of crosses might prove inadequate
as a means by which to distinguish between epigenetic and genetic events.
Because in most plants organelles are transmitted with the maternal
cytoplasm, it would be difficult to exclude the possibility that an altered
phenotype results from the influence of nongenetic components of the
cytoplasm on the expression of mitochondrial or chloroplast genes.

In contrast to the cases cited earlier, resistance to cycloheximide was
very unstable in cell lines of *N. tabacum* that had been selected for growth
in the presence of the drug (Maliga et al., 1976). Resistant cell lines were
cloned to eliminate the possibility that the rapid disappearance of cycloheximide resistance on a nonselective medium resulted from overgrowth of a heterogeneous cell population by normal cells. Colonies
capable of growth on medium supplemented with cycloheximide at 0.5
μg/ml were obtained from only one of four initially resistant cell lines that
were plated. The one clone that was studied exhibited the same behavior
as the resistant cell culture from which it had been isolated: resistance
was displayed by callus cultures as long as they were maintained in the
presence of cycloheximide, but sensitivity was restored by a single
passage in the absence of the drug. An association between cycloheximide resistance and organogenesis was suggested by the observation that resistant callus cultures formed shoots when propagated on a
medium that completely inhibited differentiation of callus that had lost
resistance to cycloheximide and differentiation of unselected control
callus. The peroxidase isozyme pattern of resistant cells cultured on
callus-maintenance medium supplemented with cycloheximide was distinct from that of sensitive cells grown on the same medium without
cycloheximide, and it resembled the banding pattern of sensitive callus
cultured on a medium that promotes organogenesis. However, if resistant
cells were cultured on callus-maintenance medium in the absence of
cycloheximide, conditions under which the resistance phenotype vanished, the isozyme pattern began to shift to one characteristic of normal

sensitive cells growing on the same medium. On the basis of this evidence, Maliga and associates (1976) concluded that resistance to cycloheximide is accomplished by the activation of genes that are expressed normally in differentiated tissues and not in callus cultures. However, since nonselected differentiating tissue is sensitive to cycloheximide (Maliga, 1978), it would appear that resistance of cell cultures was conferred by an exceptionally activated gene that normally is not expressed at either the callus stage or the early organogenetic stage or by a mutant allele of a gene that is expressed in the latter stage but not the former stage.

Epigenetic changes that occur in cultured cells may be viewed as the expression in such cells of normal alleles of genes that usually are not active at this level of differentiation. These changes may appear stable (habituation) or unstable (cycloheximide resistance). Persistence of the variant phenotype in callus cultures formed from differentiated organs may be used as an initial indication that the modification is the result of a true genetic change, but genetic analysis by sexual crosses is the only adequate proof of a genetic alteration. Because plant regeneration is not possible in many experimental systems, other criteria by which to evaluate the molecular basis of a phenotypic change have been proposed.

One category of evidence that has been regarded as proof of a mutation concerns the identification of an altered gene product. Mutationally altered proteins may be recognized by amino acid sequencing, electrophoresis of tryptic digests, thermolability properties, or kinetic experiments that determine the affinity of an enzyme for substrates, cofactors, or allosteric effectors. Thus a tobacco cell line resistant to 5-methyltryptophan was considered to be a mutant when a species of anthranilate synthetase less sensitive than the normal enzyme to feedback inhibition by tryptophan and 5-methyltryptophan was detected in crude extracts (Widholm, 1972a). However, with the discovery of both tryptophan-sensitive and -insensitive isozymes of anthranilate synthetase in cultured cells of *Solanum tuberosum* (Carlson and Widholm, 1978), the limitation of this approach became clear. In many cases it cannot be used to discriminate between the product of a mutant allele of a structural gene (genetic change) and the synthesis of an isozyme that is encoded by a distinct gene and normally is not expressed under those particular conditions (epigenetic change). This quandary was encountered again in the analysis of *S. tuberosum* cell lines resistant to 5-methyltryptophan. The two anthranilate synthetase isozymes of potato are electrophoretically separable, and whereas the major activity found in normal cells is inhibited by tryptophan, the minor activity is not. However, 5-methyltryptophan-resistant cell lines contain less of the feedback-sensitive activ-

ity and more of the insensitive activity. Resistar[...] apparently was achieved in these cells by incr[...] of the feedback-insensitive isozyme (Carlson [...] whether the altered rates of synthesis of the [...] physiological adaptation, an epigenetic event, [...]

It also has been suggested that genetic an[...] contrasted on the basis of their frequencies of [...] of the effects of mutagenesis on these fre[...] Maliga, 1976). It is reasoned that the fre[...] expression should be fairly high and insensit[...] structural changes in the DNA are expected [...] that can be increased by treatment with a m[...] events per cell generation that has been es[...] (1973) for cytokinin habituation in tobac[...] characteristic of epigenetic phenomena. H[...] quency of unstable cycloheximide resistan[...] 1976), and that of picloram resistance v[...] (Chaleff and Parsons, 1978b). Moreover, i[...] occurred more frequently, since, of seven [...] behaved as true mutations, and only one [...] epigenetic event. In addition to these [...] employing the frequency of appearance o[...] for distinguishing between genetic and [...] sider that posed by the occurrence in h[...] modifications involving transposable e[...] interfere with gene activity by inserting [...] excision can restore gene function eith[...] a stable loss of activity. By these mec[...] to induce mutations at a high frequer[...] example, transposable elements caus[...] frequencies of 0.1 to 50 percent (as [...] 1974). The existence of transposable [...] event an unacceptable standard by w[...] the high frequency of their occurr[...] transposable elements results in str[...] can be transmitted through crosses [...] genetic events.

Other objections can be raised ag[...] evaluate the nature of a phenotypic [...] loci at which mutations can occur [...] known. For example, let us supp[...] different loci, each mutating spont[...]

in a [...]
phenc[...]
transf[...]
about [...]
meani[...]
a frac[...]
the gr[...]
freque[...]
are dis[...]
that ex[...]
of app[...]
demons[...]
control[...]
reporte[...]
ized bac[...]
mutager[...]
failure [...]
epigenet[...]

This [...]
ineluctal[...]
through [...]
distingui[...]
us that v[...]
evaluatio[...]
the conv[...]
segregati[...]
effects o[...]
experime[...]
none sho[...]
conclusio[...]
provided [...]
technique[...]
improves,[...]
demand t[...]
haps epig[...]
changes t[...]
revealed. [...]
require tha[...]
which a g[...]
unequivoca[...]
altered phe[...]
genetic ana[...]
phenotype [...]

Second corollary: not all genes are expressed in the mature plant

Certain genes that are expressed in undifferentiated cultured cells may not be active in the adult plant, or at least not in all of the differentiated organs of the plant. In this circumstance a true mutational event might be visible in cell culture but not alter the phenotype of the plant. The regeneration from a variant cell line of what appears to be a perfectly normal plant might be construed as prima facie evidence that an epigenetic event rather than a genetic event is responsible for the altered behavior of the cultured cells. But to resolve this question properly it first will be necessary to inspect a callus culture formed from the regenerated plant (secondary callus). If the variant phenotype is not displayed by the secondary callus, an epigenetic basis may be assumed. However, reappearance of the variant phenotype in callus cultures derived not only from the regenerated plant but also from the progeny of that plant is convincing proof that a mutation has occurred in a gene that is expressed only in culture, not in the plant.

A glycerol-utilizing *(Gut)* mutant of *N. tabacum* provides an example of a genetic change that is visible only in cultured cells, not in the plant (Chaleff and Parsons, 1978a). *Gut* cells are able to grow on glycerol as the sole source of carbon and energy, whereas normal cells cannot. In a genetic analysis performed by testing callus cultures derived from the progeny of various crosses for the capacity to grow on glycerol, it was established that the ability to utilize glycerol is conferred by a dominant allele of a single nuclear gene. However, no morphological differences were apparent between normal plants and plants from which *Gut* callus cultures were obtained, nor were there any differences in germination between normal seeds and mutant seeds or in growth of seedlings in the dark on a glycerol medium (R. S. Chaleff, unpublished data). The problem with such experiments is that one really does not know where to look for expected differences. The *Gut* plants may well be producing an altered protein that does not affect gross morphological characteristics and that can be detected only by direct biochemical assay.

Tissue-specific expression of an altered phenotype was also observed with mutants of tobacco that were resistant to hydroxyurea. By means of monitoring of growth responses to hydroxyurea in callus cultures derived from progeny plants, it was demonstrated that in crosses the mutant allele segregated as a Mendelian dominant. However, germination of seeds from mutant plants was no less sensitive to hydroxyurea than was that of normal seeds, and mutant plants were morphologically normal (R. L. Keil and R. S. Chaleff, unpublished results). An advantage of this system is that a clue to the identity of the altered enzyme is provided by the knowledge that in mammalian cells resistance is associated with diminished sensitivity of ribonucleotide reductase to inhibition by hydroxy-urea (Lewis and Wright, 1974). Therefore, enzyme assays can be performed to identify an altered activity and to determine if it is present only in cell cultures or in the plant as well.

The occurrence of multiple genes that encode proteins with similar catalytic activities (isozymes) provides another source of potential differences in gene expression between callus and plant. If different isozymes are produced in the callus and in the plant, selection in culture for an altered activity will permit recovery of only mutant alleles of the gene that encodes the isozyme that predominates in cultured tissue. In the adult plant this gene will not be active, and the normal form of the other isozyme will be synthesized. This plant, although harboring a mutant allele, should appear perfectly normal. The most direct and conclusive test for the identity of enzyme activities produced by the plant and by the callus is comparison of their physical and chemical properties. By such comparative studies it has been shown that distinct isozymes of aspartokinase are synthesized by roots and by cultured tissue of carrot. Aspartokinase extracted from fresh carrot root tissue is inhibited greatly by threonine and only slightly by lysine. In contrast, the major activity present in root tissue slices after 3 days of culture and in well-established cell suspension cultures is relatively sensitive to lysine and insensitive to threonine (Matthews and Widholm, 1978; Sakano and Komamine, 1978). Extracts of fresh tissue and cell suspension cultures contained similar forms of two additional activities of the lysine biosynthetic pathway, homoserine dehydrogenase and dihydrodipicolinic acid synthase. Homoserine dehydrogenase activities from both sources were inhibited by threonine and cysteine, and electrophoresis of the separate and combined extracts revealed only a single band of activity. Dihydrodipicolinic acid synthase activities from root and cultured tissues were inhibited to the same degree by lysine and were electrophoretically identical (Matthews and Widholm, 1978).

A different pattern of aspartokinase regulation apparently exists in maize. Gengenbach and associates (1978) prepared extracts from several maize tissues (shoot, root, kernel, callus, and suspension culture) and analyzed the allosteric inhibition properties of the first three enzymes of the branched pathway by which lysine, methionine, threonine, and isoleucine are synthesized. The aspartokinase activities in all five extracts exhibited similar responses to various combinations of the four amino acids. The aspartate semialdehyde dehydrogenase activity present in extracts of suspension cultures was more sensitive to inhibition by methionine, and the homoserine dehydrogenase activity isolated from callus tissue was inhibited to a greater degree by threonine than were the function altered by habituation, maintenance of endogenous concentrations of hormones, is critical to differentiation and that differentiation might proceed only if normal hormone levels are restored. It is possible that other cases of epigenetic change might involve altered expression of more general cellular functions that are also active in differentiated tissue. Such changes, although epigenetic, might endure through a cycle of differentiation and dedifferentiation. Therefore, the only acceptable criterion that can be applied to distinguish genetic and epigenetic events is the inheritance of a trait by progeny produced by sexual crosses. Transmission through the gametes indicates that a trait is maintained stably through meiosis as well as through mitosis and is the best possible evidence of a structural change in the DNA. In cases of extra-chromosomal inheritance, the analysis of crosses might prove inadequate as a means by which to distinguish between epigenetic and genetic events. Because in most plants organelles are transmitted with the maternal cytoplasm, it would be difficult to exclude the possibility that an altered phenotype results from the influence of nongenetic components of the cytoplasm on the expression of mitochondrial or chloroplast genes.

In contrast to the cases cited earlier, resistance to cycloheximide was very unstable in cell lines of *N. tabacum* that had been selected for growth in the presence of the drug (Maliga et al., 1976). Resistant cell lines were cloned to eliminate the possibility that the rapid disappearance of cycloheximide resistance on a nonselective medium resulted from overgrowth of a heterogeneous cell population by normal cells. Colonies capable of growth on medium supplemented with cycloheximide at 0.5 μg/ml were obtained from only one of four initially resistant cell lines that were plated. The one clone that was studied exhibited the same behavior as the resistant cell culture from which it had been isolated: resistance was displayed by callus cultures as long as they were maintained in the presence of cycloheximide, but sensitivity was restored by a single passage in the absence of the drug. An association between cycloheximide resistance and organogenesis was suggested by the observation that resistant callus cultures formed shoots when propagated on a medium that completely inhibited differentiation of callus that had lost resistance to cycloheximide and differentiation of unselected control callus. The peroxidase isozyme pattern of resistant cells cultured on callus-maintenance medium supplemented with cycloheximide was distinct from that of sensitive cells grown on the same medium without cycloheximide, and it resembled the banding pattern of sensitive callus cultured on a medium that promotes organogenesis. However, if resistant cells were cultured on callus-maintenance medium in the absence of cycloheximide, conditions under which the resistance phenotype vanished, the isozyme pattern began to shift to one characteristic of normal sensitive cells growing on the same medium. On the basis of this evidence, Maliga and associates (1976) concluded that resistance to cycloheximide is accomplished by the activation of genes that are expressed normally in differentiated tissues and not in callus cultures. However, since nonselected differentiating tissue is sensitive to cycloheximide (Maliga, 1978), it would appear that resistance of cell cultures was conferred by an exceptionally activated gene that normally is not expressed at either the callus stage or the early organogenetic stage or by a mutant allele of a gene that is expressed in the latter stage but not the former stage.

Epigenetic changes that occur in cultured cells may be viewed as the expression in such cells of normal alleles of genes that usually are not active at this level of differentiation. These changes may appear stable (habituation) or unstable (cycloheximide resistance). Persistence of the variant phenotype in callus cultures formed from differentiated organs may be used as an initial indication that the modification is the result of a true genetic change, but genetic analysis by sexual crosses is the only adequate proof of a genetic alteration. Because plant regeneration is not possible in many experimental systems, other criteria by which to evaluate the molecular basis of a phenotypic change have been proposed.

One category of evidence that has been regarded as proof of a mutation concerns the identification of an altered gene product. Mutationally altered proteins may be recognized by amino acid sequencing, electrophoresis of tryptic digests, thermolability properties, or kinetic experiments that determine the affinity of an enzyme for substrates, cofactors, or allosteric effectors. Thus a tobacco cell line resistant to 5-methyltryptophan was considered to be a mutant when a species of anthranilate synthetase less sensitive than the normal enzyme to feedback inhibition by tryptophan and 5-methyltryptophan was detected in crude extracts (Widholm, 1972a). However, with the discovery of both tryptophan-sensitive and -insensitive isozymes of anthranilate synthetase in cultured cells of *Solanum tuberosum* (Carlson and Widholm, 1978), the limitation of this approach became clear. In many cases it cannot be used to discriminate between the product of a mutant allele of a structural gene (genetic change) and the synthesis of an isozyme that is encoded by a distinct gene and normally is not expressed under those particular conditions (epigenetic change). This quandary was encountered again in the analysis of *S. tuberosum* cell lines resistant to 5-methyltryptophan. The two anthranilate synthetase isozymes of potato are electrophoretically separable, and whereas the major activity found in normal cells is inhibited by tryptophan, the minor activity is not. However, 5-methyltryptophan-resistant cell lines contain less of the feedback-sensitive activity and more of the insensitive activity. Resistance to 5-methyltryptophan apparently was achieved in these cells by increasing the relative amount of the feedback-insensitive isozyme (Carlson and Widholm, 1978), but whether the altered rates of synthesis of the two isozymes result from physiological adaptation, an epigenetic event, or mutation is unknown.

It also has been suggested that genetic and epigenetic events can be contrasted on the basis of their frequencies of occurrence and on the basis of the effects of mutagenesis on these frequencies (Widholm, 1974a; Maliga, 1976). It is reasoned that the frequency of changes in gene expression should be fairly high and insensitive to mutagenesis, whereas structural changes in the DNA are expected to occur at a low frequency that can be increased by treatment with a mutagen. Thus the rate of 10^{-3} events per cell generation that has been estimated by Binns and Meins (1973) for cytokinin habituation in tobacco cells is presumed to be characteristic of epigenetic phenomena. However, the spontaneous frequency of unstable cycloheximide resistance was 4×10^{-6} (Maliga et al., 1976), and that of picloram resistance was estimated to be 2×10^{-5} (Chaleff and Parsons, 1978b). Moreover, in the latter case genetic events occurred more frequently, since, of seven resistant cell lines selected, four behaved as true mutations, and only one could have been caused by an epigenetic event. In addition to these inconsistencies that arise from employing the frequency of appearance of an altered phenotype as a basis for distinguishing between genetic and epigenetic events, we must consider that posed by the occurrence in higher plants of a class of genetic modifications involving transposable elements. Transposable elements interfere with gene activity by inserting in a gene or close to a gene. Their excision can restore gene function either fully or partially or can produce a stable loss of activity. By these mechanisms transposable elements act to induce mutations at a high frequency in many plants. In maize, for example, transposable elements cause mutations in the germ cells at frequencies of 0.1 to 50 percent (as reviewed by Fincham and Sastry, 1974). The existence of transposable elements makes the frequency of an event an unacceptable standard by which to determine its cause. Despite the high frequency of their occurrence, the insertion or excision of transposable elements results in structural alterations to the DNA that can be transmitted through crosses and therefore must be classified as genetic events.

Other objections can be raised against the use of rates or frequencies to evaluate the nature of a phenotypic change. First, the number of different loci at which mutations can occur to produce a certain phenotype is not known. For example, let us suppose that a mutation at any one of ten different loci, each mutating spontaneously at a rate of 10^{-5}, could result in a given phenotypic change. The rate of appearance of that new phenotype then would be 10^{-4}, which might persuade some that the transformation is epigenetic. Thus, in the absence of any information about the number of loci that can be involved, rates and frequencies are meaningless. Interpreting such data would be like trying to make sense of a fraction without knowing the denominator. Second, several features of the growth of plant cells in culture make estimation of the rate or frequency of phenotypic change problematic. These technical difficulties are discussed in Chapter 4. However, it should be mentioned en passant that experiments designed to determine the effect of a mutagen on the rate of appearance of a given phenotype should include a control that demonstrates the potency of the mutagen. To my knowledge, this simple control has not been performed in any such experiment that has been reported. The effect of the mutagen on the mutation rate of a characterized bacterial or yeast gene would serve as an easy and direct test of its mutagenic activity. Otherwise, of course, one does not know whether the failure of a treatment to effect an increase in rate results from the epigenetic nature of the change or from the inactivity of the mutagen.

This cursory consideration of genetic and epigenetic events leads us ineluctably to the conclusion that at the present time, transmission through sexual crosses is the sole valid criterion by which they can be distinguished. But the diversity of biological systems should have taught us that we should not rush to embrace any absolute standards for their evaluation. The strength of formal genetic analysis is that it is based on the convergence of many different lines of experimentation: predicted segregation patterns from crosses, mutation and reversion frequencies, effects of mutagens, recombination, and even DNA sequencing. One experimental approach may be more convenient in a given system, but none should be singled out as having more merit than another. Instead, conclusions must be derived from and must be consistent with evidence provided by several types of experiments. We must also remember that as techniques are refined and the resolution of our analytical methods improves, new discoveries will be made that will change our concepts and demand that definitions and standards be rewritten or abandoned. Perhaps epigenetic events that are transmitted through meiosis and genetic changes that are not visible in subsequent generations may yet be revealed. Nevertheless, for the moment, an operational definition would require that the term ''mutant'' be reserved exclusively for those cases in which a genetic basis for an altered phenotype has been established unequivocally by studies of inheritance across sexual generations. If an altered phenotype is not heritable (in which case it is epigenetic) or if a genetic analysis has not been performed, cell lines exhibiting that phenotype should be referred to as ''variants.''

Second corollary: not all genes are expressed in the mature plant

Certain genes that are expressed in undifferentiated cultured cells may not be active in the adult plant, or at least not in all of the differentiated organs of the plant. In this circumstance a true mutational event might be visible in cell culture but not alter the phenotype of the plant. The regeneration from a variant cell line of what appears to be a perfectly normal plant might be construed as prima facie evidence that an epigenetic event rather than a genetic event is responsible for the altered behavior of the cultured cells. But to resolve this question properly it first will be necessary to inspect a callus culture formed from the regenerated plant (secondary callus). If the variant phenotype is not displayed by the secondary callus, an epigenetic basis may be assumed. However, reappearance of the variant phenotype in callus cultures derived not only from the regenerated plant but also from the progeny of that plant is convincing proof that a mutation has occurred in a gene that is expressed only in culture, not in the plant.

A glycerol-utilizing *(Gut)* mutant of *N. tabacum* provides an example of a genetic change that is visible only in cultured cells, not in the plant (Chaleff and Parsons, 1978a). *Gut* cells are able to grow on glycerol as the sole source of carbon and energy, whereas normal cells cannot. In a genetic analysis performed by testing callus cultures derived from the progeny of various crosses for the capacity to grow on glycerol, it was established that the ability to utilize glycerol is conferred by a dominant allele of a single nuclear gene. However, no morphological differences were apparent between normal plants and plants from which *Gut* callus cultures were obtained, nor were there any differences in germination between normal seeds and mutant seeds or in growth of seedlings in the dark on a glycerol medium (R. S. Chaleff, unpublished data). The problem with such experiments is that one really does not know where to look for expected differences. The *Gut* plants may well be producing an altered protein that does not affect gross morphological characteristics and that can be detected only by direct biochemical assay.

Tissue-specific expression of an altered phenotype was also observed with mutants of tobacco that were resistant to hydroxyurea. By means of monitoring of growth responses to hydroxyurea in callus cultures derived from progeny plants, it was demonstrated that in crosses the mutant allele segregated as a Mendelian dominant. However, germination of seeds from mutant plants was no less sensitive to hydroxyurea than was that of normal seeds, and mutant plants were morphologically normal (R. L. Keil and R. S. Chaleff, unpublished results). An advantage of this system is that a clue to the identity of the altered enzyme is provided by the knowledge that in mammalian cells resistance is associated with diminished sensitivity of ribonucleotide reductase to inhibition by hydroxy-urea (Lewis and Wright, 1974). Therefore, enzyme assays can be performed to identify an altered activity and to determine if it is present only in cell cultures or in the plant as well.

The occurrence of multiple genes that encode proteins with similar catalytic activities (isozymes) provides another source of potential differences in gene expression between callus and plant. If different isozymes are produced in the callus and in the plant, selection in culture for an altered activity will permit recovery of only mutant alleles of the gene that encodes the isozyme that predominates in cultured tissue. In the adult plant this gene will not be active, and the normal form of the other isozyme will be synthesized. This plant, although harboring a mutant allele, should appear perfectly normal. The most direct and conclusive test for the identity of enzyme activities produced by the plant and by the callus is comparison of their physical and chemical properties. By such comparative studies it has been shown that distinct isozymes of aspartokinase are synthesized by roots and by cultured tissue of carrot. Aspartokinase extracted from fresh carrot root tissue is inhibited greatly by threonine and only slightly by lysine. In contrast, the major activity present in root tissue slices after 3 days of culture and in well-established cell suspension cultures is relatively sensitive to lysine and insensitive to threonine (Matthews and Widholm, 1978; Sakano and Komamine, 1978). Extracts of fresh tissue and cell suspension cultures contained similar forms of two additional activities of the lysine biosynthetic pathway, homoserine dehydrogenase and dihydrodipicolinic acid synthase. Homoserine dehydrogenase activities from both sources were inhibited by threonine and cysteine, and electrophoresis of the separate and combined extracts revealed only a single band of activity. Dihydrodipicolinic acid synthase activities from root and cultured tissues were inhibited to the same degree by lysine and were electrophoretically identical (Matthews and Widholm, 1978).

A different pattern of aspartokinase regulation apparently exists in maize. Gengenbach and associates (1978) prepared extracts from several maize tissues (shoot, root, kernel, callus, and suspension culture) and analyzed the allosteric inhibition properties of the first three enzymes of the branched pathway by which lysine, methionine, threonine, and isoleucine are synthesized. The aspartokinase activities in all five extracts exhibited similar responses to various combinations of the four amino acids. The aspartate semialdehyde dehydrogenase activity present in extracts of suspension cultures was more sensitive to inhibition by methionine, and the homoserine dehydrogenase activity isolated from callus tissue was inhibited to a greater degree by threonine than were the activities from other sources. However, because the cultured and adult plant tissues used in these experiments originated from different cultivars, clear interpretation of these results is not possible. Any disparity between the activities in the extracts of these maize tissues could be due to genotypic differences rather than to synthesis of tissue-specific isozymes. Electrophoretic separation of several homoserine dehydrogenase isozymes in extracts of callus and suspension cultures of the same cultivar proved more meaningful. A threonine-insensitive activity present in the suspension cultures was not detected in extracts of callus tissue. Although this experiment indicated that different forms of homoserine dehydrogenase are made in the two tissues, the numbers of genes contributing to these forms are not known.

A more detailed analysis was performed by Polacco and Havir (1979), who studied the properties of urease activity of seeds and of cell suspension cultures of soybean. These studies compared activity of purified seed enzyme with activity in crude and partially purified extracts of cell suspensions by gel filtration, electrophoresis, and reactivity with an antibody preparation specific to soybean seed urease from which all antibodies that cross-react with jackbean urease had been removed. In addition to a urease that is indistinguishable from the seed enzyme by these criteria, crude extracts of cell suspensions contained other forms of urease. However, since the seeds and the cell cultures were obtained from different soybean cultivars, again we are left wondering about the nature and origin of the additional activity produced by the cell cultures.

Considered collectively, these experiments suggest that both unique and common species of enzymes are made by different tissues of several plants. These variant forms could result from aggregation of a single basic subunit or other alterations of the protein itself or from de novo synthesis of other isozymes. However, it is clear that there is no single universal answer to the question whether or not an enzyme activity present in the plant and that produced in callus cultures are in fact encoded by the same gene. In each individual system it will be necessary to resolve this question by independent experiment.

It is also to be anticipated that certain genes, although expressed in culture, are dispensable for the cultured cell but are essential to certain processes of the whole plant. Such a situation could be exploited to reveal the critical roles played by particular enzymes during development or in the adult plant. It was in this manner that Müller and Grafe (1978) recently suggested a requirement for nitrate reductase in the formation of shoots by tobacco callus. They found that organogenesis could be promoted in mutants that retained residual nitrate reductase activity, but not in mutants that completely lacked this activity. The capability to produce shoots was restored in hybrids formed by fusion of protoplasts of two complementing nonleaky mutants (Glimelius et al., 1978).

Not all genes are expressed in the cultured cell

The converse of the second corollary is equally valid: certain genes that are expressed in the adult plant may not be active in cultured cells. Clearly, since altered function cannot be selected in the absence of gene expression, it will not be possible to use cell cultures to identify mutant forms of such genes. As many differentiated tissues and organs of the adult plant (e.g., leaves, flowers, seeds) are not present in cell cultures, it is to be expected that the specialized molecules that contribute to the structure and function of those tissues and organs also will be absent from cultured cells and that the genes controlling the synthesis of such molecules will be inactive. Thus one cannot hope to select in culture for altered root morphology or for different numbers of stomata. Similarly, alterations in traits that are produced by the integration of several structures and functions of the whole plant cannot be selected among cultured cells. Yield and pest resistance are two such traits. Thus a cell culture system imposes the most severe experimental constraint of precluding selection for changes in anything other than basic cellular functions. The actual limitations encountered will depend on the species and the agent and method of selection. For example, Chaleff and Parsons (1978b) found that sensitivity of tobacco to the herbicide picloram is expressed by cultured cells, and they were able to select directly at this level for resistance that is manifested by the whole plant. However, Radin and Carlson (1978) found tobacco cell cultures to be unaffected by two herbicides (Bentazone and Phenmedipharm) to which the plant is very susceptible.

The equal sensitivities to NaCl of callus cultures of the halophyte glasswort (*Salicornia*) and callus cultures of cabbage, sweet clover, and sorghum suggest that salt tolerance is accomplished in the adult plant by a mechanism that relies on differentiated structures and specialized processes present in the whole plant but not in cultured cells (Strogonov, 1970). Nevertheless, NaCl tolerance has been selected in cultured cells of several species (see Chapter 4). Presumably, salt tolerance is achieved in these cells by a different means than in *Salicornia*. In this case, although a discrepancy between the behavior of the plant and the behavior of the cultured cells initially seemed discouraging, a real possibility exists of selecting in vitro for a mutation that introduces a mechanism different from (but as effective as) the mechanism of salt tolerance that has evolved in certain plants.

Studies of the resistance of plants to virus infections have revealed that this phenotype can be effected by several very different mechanisms, some of which function exclusively in vivo and some of which are also expressed in vitro. Of 1031 varieties of *Vigna sinensis* (cowpea) that were surveyed by inoculating leaves of seedlings, 65 proved operationally immune to the SB isolate of cowpea mosaic virus (CPMV-SB). Protoplasts from 55 of these immune varieties were then innoculated with virus in vitro, and after an incubation period the numbers of virus particles contained within the protoplasts were determined. Significant virus multiplication occurred in protoplasts of all but 1 of the 55 varieties that were immune as seedlings (Beier et al., 1977). Only a very small and delayed increase in virus titer could be detected in protoplasts of this one resistant cowpea variety (Beier et al., 1979).

Two types of resistance mechanisms also have been distinguished in the responses of tomato leaf discs and protoplasts to infection with a tomato strain of tobacco mosaic virus (TMV-L). These experiments employed isogenic tomato lines that differed from the susceptible normal variety only at either or both of two loci by alleles conferring resistance to TMV-L infection. Virus multiplication did not occur in leaf discs excised from plants homozygous for *Tm-2* or *Tm-2²* (both alleles of the same gene), but it did occur in protoplasts isolated from these plants. In contrast, no accumulation of virus antigen or increase in infectivity was observed in either leaf discs or protoplasts obtained from a plant homozygous for *Tm-1* (Motoyoshi and Oshima, 1975, 1977). Thus a given phenotype of the whole plant may be produced by any of several means. Some of these mechanisms may operate only in the intact plant, but others may be manifested at the cellular level as well. This knowledge should militate against any tendency inspired by study of but a single case to dismiss cell culture as being inapplicable to a particular problem.

For the moment, many whole plant traits still may appear inaccessible to cell culture technology. However, future developments could well bring such traits within reach of the culture vessel. Such advances might be accomplished through discovery of associations between important plant characteristics and activities that are expressed in culture. It may then prove possible to produce alterations in whole plant characteristics by selecting in culture for mutations affecting the correlated activity. Another approach would be through identification of specific biochemical pathways that contribute in some way to what is ostensibly an exclusive feature of the whole plant. It can be envisioned, for example, that resistance to lodging may be improved by selection in culture for overproduction of lignin or that a plant displaying greater resistance to insects could be regenerated from a cell line that had been selected for a higher rate of alkaloid synthesis.

3

Origins of variability in cultured plant cells

Variability in plant cell cultures

An important premise that is accepted as axiomatic for most genetic systems is that spontaneous variation arises at a low frequency. Thus, in analyzing a variant isolated from an unmutagenized population, one assumes generally that the genome is unaltered but for a single modification that is responsible for the new phenotype. But does this same principle apply to cultured plant cells?

Let us consider a cell line initiated from a tissue explant. After propagating this cell line in vitro for many cell generations, we will isolate from it clones of single-cell origin. More than likely we shall discover that a number of the clones differ from one another and from the parental cell line in characteristics such as color, morphogenetic capacity, and degree of friability. Similarly, if we were to regenerate plants from a population of cultured cells, we would find many of these plants phenotypically distinct from the plant from which the cell cultures derived. Variations of this sort are observed commonly in cell cultures of many species, and as far as we can tell, they are random and unselected. As such, these phenotypic alterations will be considered separately from those rarer modifications that can be recovered only by deliberate selection (see Chapter 4).

The literature is replete with reports documenting the variability of cultured cells and of regenerated plants. It is not my intention or desire to attempt to review these many examples, as this task has recently been done so well by D'Amato (1977, 1978), Sunderland (1977), and Skirvin (1978). But I shall bring forward a few illustrations.

In one particularly thorough study Lutz (1977) isolated more than 400 clones from a habituated (See Chapter 2) tobacco cell line. Considerable diversity of several visible features was found among these cloned cell lines. They also differed in their abilities to differentiate. Some clones could not be induced to form plants under any conditions. Plant regeneration could be promoted from other clones by the addition of hormones to the medium, and in others, plant formation occurred spontaneously. In total, only 16 of the cloned cell lines proved capable of producing plants.

Plants regenerated from many of these clones were morphologically abnormal. Moreover, the altered plant phenotype was peculiar to a given clone, and in most cases it had become a stable characteristic of that clone. Thus all plants regenerated from an individual clone had the same appearance, but plants derived from another clone manifested a different anomaly.

Variability has also been revealed in studies of sugarcane cell cultures. Five cloned cell lines originating from single cells isolated from callus of one sugarcane cultivar differed in color and growth rate (Nickell and Maretzki, 1972). In addition, many morphological variants were present among plants regenerated from callus cultures. A dwarf plant possessing an altered esterase isozyme banding pattern was produced from one callus culture (Heinz and Mee, 1969). In a subsequent experiment, regenerated plants differed from the parental variety in auricle length, pubescence, and the isozyme banding patterns of peroxidase, amylase, and transaminase activities, although among these plants no variation was observed in the esterase isozyme pattern. These biochemical and morphological differences were not correlated: altered isozyme patterns were found in some morphologically normal plants and not in all morphological variants (Heinz and Mee, 1971).

As variations have been observed among regenerated sugarcane plants for all properties thus far chosen for characterization, one may now begin to wonder what other modifications may be discovered in such plants. Of course, attention is first directed to traits of agronomic importance. Heinz (1973) screened for resistance to eyespot disease by treating diploid sugarcane plants regenerated from unmutagenized callus and suspension cultures with the toxin (helminthosporoside) that is elaborated by *Helminthosporium sacchari,* the causative agent of the disease. Quite surprisingly, 15 to 20 percent of the plants proved resistant. Furthermore, the frequency of disease resistance among regenerated plants was not increased by exposing the cell cultures to a chemical mutagen (methyl methanesulfonate) or ionizing radiation (Heinz et al., 1977). Similarly, Krishnamurthi and Tlaskal (1974) reported resistance to Fiji disease in 4 of 38 plants that had been regenerated from callus of a susceptible sugarcane variety. This resistance has been maintained through several generations of vegetative propagation.

More recently, an analogous experiment was performed with potato. Variation (presumably morphological) apparent among plants regenerated from potato mesophyll protoplasts suggested to Matern, Strobel, and Shepard (1978) that differences might also be discovered among such plants in other less immediately visible but more valuable traits. A population of 500 protoplast-derived plants was inoculated with a partially purified toxin preparation extracted from cultures of *Alternaria solani* to identify individuals resistant to early blight disease. The toxin failed to elicit the development of characteristic lesions on two plants. These plants also proved resistant to infection with the fungus. Furthermore, this resistance was stable insofar as it continued to be expressed in a second vegetative generation grown from tubers of these two plants.

Although the studies cited serve to portray the phenomenon of variability among cultured cells and regenerated plants, they have in common two features that limit their contribution to our understanding of the source and nature of such variability. First, the regenerated plants were not analyzed genetically. Now we can be fairly certain that the cause of the alterations was not merely physiological, because in the case of tobacco the anomalies of the regenerated plants were peculiar to each clone, rather than being of one kind, and in the cases of sugarcane and potato, expression persisted through a second vegetative generation. But without the results of sexual crosses we cannot distinguish between an epigenetic basis and a genetic basis for these changes (see Chapter 2). Second, even if the modifications could be shown to be genetic, we might not be closer to an understanding of their origins. If the new phenotype results from a point mutation, the chromosome number of the altered cell line or regenerated plant might be unchanged. But the high frequency of these alterations suggests that a more common event, such as a change in ploidy, might be responsible. Of course, such differences in chromosome number can be expected to contribute to phenotypic alterations in both cultured cells and regenerated plants. But because all three plants species that we have considered thus far have been polysomatic, we do not know whether such karyotypic changes arose in culture or were present beforehand in the cells of the plant from which the culture was derived. Polysomatic plants are composed of cells of different ploidies that occur as a result of chromosome replication in the absence of cell and nuclear division (endoreduplication) during the normal process of differentiation. In such plants the diploid chromosome complement is stringently preserved only in cells of the meristems and the germ line. Because explants of polysomatic species can comprise cells of different ploidies, heterogeneity of derivative cell cultures can arise by multiplication of cells of different karyotypes present within the tissue explant as well as by endomitosis or endoreduplication in vitro after the cell culture has been established. For example, it has been shown that when tobacco pith (Patau and Das, 1961) and pea root (Torrey 1965; Matthysse and Torrey, 1967) explants are placed in culture, polyploid cells present in the explants enter into mitosis. But do not despair. Several experiments have revealed something of how cells respond to being torn from their ordered realm within the plant and plunged into the anarchy of the culture vessel.

Genetic variability in cultured cells

Cytological studies

The most direct evidence that abundant genetic changes occur during proliferation of cells in culture is furnished by nuclear cytology. Variations in number and structure of chromosomes are visible proof of genetic change. As already mentioned, polysomatic plants are unsatisfactory material for studying the generation of karyotypic instability in vitro because one cannot discern whether polyploid cells originated from endoreduplicated cells present in the explant or originated by anomalous cell divisions that are a peculiar feature of the culture system itself. But there are ways of circumventing this uncertainty.

First, cultures can be initiated from cells of known chromosome number. Then only in vitro events can contribute to altering the chromosome complement of progeny cells or regenerated plants. Thus, Libbenga and Torrey (1973) have demonstrated that diploid pea root cortex cells undergo chromosome endoreduplication when placed in a medium containing auxin and cytokinin. Single cells of a specified ploidy that can be stimulated to divide in vitro are also available from pollen culture and from certain anther culture systems in which embryogenesis or callus formation has been shown to proceed from the immature pollen contained within the cultured anther rather than from the surrounding somatic tissue. The recovery of polyploid plants from cultured anthers of *Oryza sativa* (Nishi and Mitsuoka, 1969; Niizeki and Oono, 1971; Oono, 1975, 1978), *Datura innoxia* (Engvild, Linde-Laursen, and Lundqvist, 1972; Sunderland, Collins, and Dunwell, 1974), and *Nicotiana tabacum* (Kasperbauer and Collins, 1972) is clear evidence of the occurrence of abnormal mitoses in vitro. Polyploidization of pollen-derived cell cultures and plants most likely results from endomitosis and from nuclear fusion in multinucleate cells (Sunderland, 1973).

As the karyotype of the initial cell is not often known, a second procedure that can be used to reveal sources of variability in culture is to clone single cells of an established culture. If variability is not generated during divisions in vitro, all members of the cloned population will be identical. But any heterogeneity in that population must have originated in culture. Because cloning permits monitoring the changes arising from divisions of a single cell, the ploidy of the initial cell and that it may have come from a polysomatic plant are unimportant. To analyze karyotypic variability occurring in culture, Murashige and Nakano (1967) cloned cells from callus derived from tobacco pith. The variation in chromosome numbers and the high frequency of aneuploidy that appeared in many of the cloned cell lines only could have been produced in vitro.

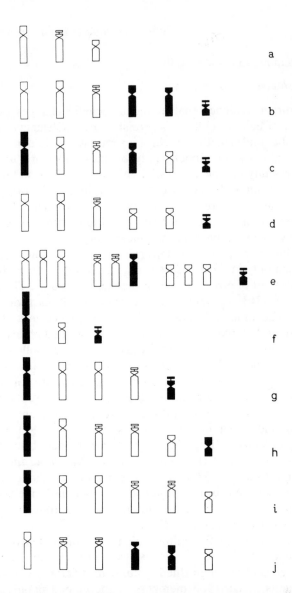

FIGURE 3.1. Idiograms depicting karyotypes of cultured cells of *Crepis capillaris*. *a:* Normal haploid with (from left to right) one long chromosome with subterminal centromere (L), one chromosome with satellite region (SAT), and one short chromosome with subterminal centromere (S). *b–e:* Abnormal karyotypes observed in callus cultures derived from a diploid plant. *b:* Diploid with translocations involving one SAT chromosome and both S chromosomes. *c:* Diploid in which rearrangements involving all three chromosome pairs have occurred. *d:* Diploid with a deficient SAT chromosome. *e:* Hypotetraploid (4*n* − 2) lacking one L and one SAT chromosome and with a translocation between one SAT and one S chromosome. *f–j:* Abnormal karyotypes observed in callus cultures derived from a haploid plant. *f:* Haploid with a translocation between one L and one S

Cultures of nonpolysomatic plants provide a third means of studying the origin of variability in vitro. Because all cells of a nonpolysomatic plant have the same number of chromosomes, except for rare aberrations occurring in vitro, deviations from that number are caused by events occurring in culture. Thus observed changes in chromosome numbers of cultured cells of *Daucus carota* (Muir, 1965; Smith and Street, 1974; Bayliss, 1975) and *Crepis capillaris* (Sacristán, 1971), both of which are nonpolysomatic plants, have furnished additional documentation of the occurrence of mitotic irregularities in vitro.

The studies with *C. capillaris* also compared the rates of ploidy change in cell lines initiated from haploid and diploid plants. The rates of formation of tetraploid cells from diploid cells were approximately the same in haploid- and diploid-derived cell lines (although, because it is not clear that the plants used in these studies were isogenic, the significance of this comparison is questionable). However, within the haploid-derived cell line the rate of diploidization was much higher than the rate of tetraploidization. These results suggest that the haploid genome is relatively less stable than are the genomes of higher ploidies (Sacristán, 1971).

Another advantage of using *C. capillaris* for cytological analyses derives from the arrangement of its genome as a small number ($2n = 6$) of chromosomes that can be identified on the basis of their lengths and the constrictions at the positions of the centromeres and nucleolar organizers. These distinct morphological features make possible the detection of structural rearrangements of the chromosomes. Such rearrangements would go unnoticed in a study of a plant possessing a larger complement of indistinguishable chromosomes, where, by necessity, one would be limited merely to surveying the chromosome numbers of cultured cells. Diagrammatic representations of some of the abnormal karyotypes (idiograms) observed among cell cultures initiated from diploid (Figure 3.1 *b–e*) and haploid (Figure 3.1 *f–j*) plants illustrate chromosome loss (aneuploidy) and mutation (translocations and deficiencies).

The detection of aneuploid cells in the *C. capillaris* cultures indicates that abnormalities other than endomitosis, endoreduplication, and division of two nuclei on a common spindle occur in culture. Several studies have identified mechanisms capable of generating aneuploidy in cultured cells. Bayliss (1973) recorded mitotic spindle abnormalities (as shown by

Caption to Figure 3.1 (*cont'd*)
chromosome. *g:* Aneuploid ($2n - 1$) with rearrangements between SAT and S chromosomes. *h:* Diploid with a translocation between one L and one S chromosome. *i:* Diploid with rearrangements between one (extra) L chromosome and one S chromosome (contains more chromatin than normal diploid). *j:* Diploid with a translocation between one L and one S chromosome. Abnormal chromosomes are shown in black. (From Sacristán, 1971.)

multipolar spindle formations at anaphase), lagging chromosomes, and chromosome bridges in carrot cells from suspension cultures. In addition, dicentric chromosomes leading to bridge formation during anaphase and consequent chromosome breakage and rearrangement have been observed in cultured cells of carrot (Mitra, Mapes, and Steward, 1960), *Haplopappus gracilis* (Marks and Sunderland, 1966), *Lolium perenne* (Norstog, Wall, and Howland, 1969), and *Triticum monococcum* (Kao et al., 1970). The reader interested in a more detailed description of these processes is referred to the review by Sunderland (1977).

Although in cultured cells karyotypic instability seems to be commonplace, many cases have been reported in which the initial ploidy of a cell line is maintained. In other cases the chromosome number changes from that of the progenitor cell, but the aneuploid or polyploid condition then becomes stable. Predominance of a particular karyotype in a cultured cell line probably indicates that that karyotype is best suited to cell multiplication in vitro. Thus the preponderance of a karyotype is established by virtue of its competitive advantage over the karyotype of the initial cell from which the cell line derived and over other karyotypes that might be generated by subsequent mitotic irregularities. Karyotypic stability generally is inferred from the constancy of the chromosome number. However, structural rearrangements can continue to occur without altering the number of chromosomes. Therefore, although karyotypic stability indubitably occurs, any presumption of its establishment should be made cautiously.

Genetic analysis of regenerated plants

If one is not fond of peering through a microscope, evidence of genetic variability arising in culture can be found elsewhere. By accepting the criterion for genetic change put forth in Chapter 2 (viz., a conventional pattern of inheritance of an altered phenotype), proof of mutation may be obtained by examining progeny of plants regenerated from cell and callus cultures for increased frequencies of variant forms. However, it should be remembered that because the cytoplasm of the female parent is transmitted to the zygote, crosses in which the regenerated plant serves as the male parent provide stronger evidence of gene mutation than does self-fertilization.

Considerable variation has been reported among progeny produced by selfing rice plants regenerated from diploid callus cultures. Variations have been found in chlorophyll content, time of seed set and maturation, plant height, morphology, and fertility. Because these differences from the parent cultivar continue to be displayed in the second generation as well, they most probably are of mutational origin (Oono, 1978).

Genetic variability has also appeared among tobacco plants regenerated from a picloram-resistant cell line. Before these plants were crossed with plants regenerated from independently isolated hydroxyurea-resistant (*HuR*) cell lines (R. L. Keil and R. S. Chaleff, unpublished results) to investigate linkage of the two traits, callus cultures derived from the picloram-resistant (*PmR1*) plants were tested for sensitivity to hydroxyurea. This routine control was performed to ensure that the *PmR1* allele did not interfere with the expression of sensitivity to hydroxyurea. We were somewhat surprised to discover that the *PmR1* cell cultures were resistant to hydroxyurea. Because the original parental cell lines from which the *PmR1* mutants were selected and all other normal cell lines tested proved sensitive to hydroxyurea, hydroxyurea resistance was initially rationalized as due to a pleiotropic effect of the *PmR1* allele. This interpretation seemed to receive support from the fact that in independent experiments approximately half of the cell lines selected for picloram resistance were resistant to hydroxyurea. But our complacency was dispelled by the revelation that callus cultures derived from several picloram-sensitive isolates produced by the self-fertilization of a heterozygous *PmR1/+* plant were also resistant to hydroxyurea. This result was the first indication of the genetic independence of the mutations conferring resistance to the two antimetabolites (R. S. Chaleff and R. L. Keil, unpublished results). Accordingly, the mutation conferring resistance to hydroxyurea that was recovered in *PmR1* was designated *HuRA*. Data confirming that *HuRA* and *PmR1* are not linked were obtained from more extensive genetic analysis of a plant heterozygous for both resistance markers and for a third allele that enables callus cultures to utilize glycerol as carbon source (*PmR1/+; HuRA/+; Gut/+*). By examination of the phenotype of callus cultures derived from haploid individuals produced by culturing anthers of this plant, segregation was monitored directly among the gametes. Growth tests were also performed on callus cultures initiated from progeny of a test cross of the triply heterozygous plant with a normal plant. Although the numbers were inevitably small, the results of both studies are adequate to show all three markers to be unlinked (Table 3.1) (R. S. Chaleff and R. L. Keil, unpublished results). Separate mutations also are responsible for the picloram and hydroxyurea resistance of another isolate that was selected for resistance to picloram (*PmR7*). Self-fertilization of a doubly heterozygous (*PmR7/+; HuR/+*) plant regenerated from this cell line produced 32 *PmR; HuR*, 9 *PmR; +*, 6 +; *HuR*, and 5 +; + individuals, which is in close agreement with the theoretical ratio of 9:3:3:1 expected of two independently segregating dominant mutations. The mutation responsible for the hydroxyurea resistance of mutant *PmR7* is not linked to *HuRA* and has been denoted *HuRG* (R. S. Chaleff and R. L. Keil, unpublished results).

Table 3.1. *Independent segregation of* PmR1, HuRA, *and* Gut *alleles among progeny of a triply heterozygous tobacco plant*

	No. of individuals of phenotype								
	PmR; HuR; Gut	*+; +; +*	*PmR; +; +*	*+; HuR; Gut*	*PmR; HuR; +*	*+; HuR; +*	*PmR; +; Gut*	*+; +; Gut*	Total
Haploid plants obtained from anther culture	14	8	10	12	4	9	8	2	67
Progeny of cross *PmR1/+; HuRA/+; Gut/+ × +/+; +/+; +/+*	2	7	1	4	7	5	5	5	36

Note: If all three markers segregate independently, 8.4 individuals of each phenotype will be expected among 67 haploid plants produced by anther culture, and 4.5 individuals of each phenotype will be expected among 36 progeny of a test cross.
Source: R. S. Chaleff (unpublished data).

In contrast to the genetic independence of resistance to hydroxyurea and picloram in *PmR1* and *PmR7*, these two resistances proved inseparable by recombination among 23 progeny of another picloram-resistant mutant, *PmR6*, which is not linked to *PmR1* (Chaleff, 1980a; R. S. Chaleff and R. L. Keil, unpublished results). A relationship between resistances to the two antimetabolites was becoming evident. But the nature of this association was somewhat of a mystery because of the genetic independence of the *HuR* mutations discovered in the *PmR1* and *PmR7* mutants. Perhaps the *HuR* alleles enhanced the resistance to picloram conferred by the *PmR1* and *PmR7* mutations, and therefore propagation of the *PmR1* and *PmR7* mutant cell lines in the presence of picloram exerted a selective pressure for this second mutational event. This possibility was explored by selfing the doubly heterozygous regenerated plant to obtain the single mutant *PmR7; +* and the double mutant *PmR7; HuRG* and conducting growth tests of the derivative callus cultures. Similarly, growth tests were performed with callus cultures initiated from several of the haploid plants that had been produced by culturing anthers of the triple heterozygote *PmR1/+; HuRA/+; Gut/+*. The *HuRA* mutation by itself does not afford any detectable resistance to picloram. Nor is any difference apparent in the degrees of picloram resistance of the single and double mutants (Table 3.2). However, these growth tests may not be sufficiently sensitive to detect a slight growth advantage of the double mutant. Hence, we are still left puzzling over the origin of the *HuR* mutations in the *PmR* cell lines.

The discovery in the *PmR1* and *PmR7* mutants of genetically independent mutations conferring hydroxyurea resistance offers an example of the surprises that the plant genome holds in store for us. Testing the *PmR* mutants for hydroxyurea resistance was fortuitous in that it was only a control to be performed prior to crossing them with hydroxyurea-resistant mutants. Who knows what other changes lie undetected in the genome of the cultured plant cell?

Physiological variability

It is well known that the phenotype of an individual is not merely the realization of the direct and preprogrammed expression of its genotype. Rather, the phenotype results from the interaction between genetic potential and environmental factors that can modify the activity of particular genes and their products. It stands to reason, therefore, that changes or perturbations of the environment might elicit an altered phenotype. And despite determined efforts of many scientists to reconstruct in vitro the nutritional, hormonal, and even atmospheric conditions that prevail in situ, the culture vessel represents a very alien milieu for the plant cell. This discrepancy between plant and Petri dish widens when one

Table 3.2. *Growth responses to picloram of callus cultures derived from normal and mutant plants*

Phenotype of culture	Control (no picloram) Weight (mg)	n	500 μM picloram Weight (mg)	n	Percent of control
Normal (+/ +)	526 ± 55	5	121 ± 7	5	23
HuR9[a]	625 ± 25	5	100 ± 11	5	16
Gut	1006 ± 66	31	163 ± 17	21	16

Callus cultures
initiated from diploid progeny of regenerated plant
PmR7/+; HuRG/+:

PmR7; HuRG (plant 1)	1420 ± 83	5	923 ± 117	5	65
PmR7; HuRG (plant 2)	1188 ± 81	5	869 ± 48	4	73
PmR7; + (plant 1)	921 ± 104	5	534 ± 15	5	58
PmR7; + (plant 2)	821 ± 32	5	577 ± 45	3	70

Callus cultures
initiated from haploid plants produced by culturing anthers of
PmR1/+; HuRA/+; Gut/+:

+; HuRA; Gut	1221 ± 67	31	122 ± 7	20	10
PmR1; HuRA; Gut	1236 ± 29	45	492 ± 32	23	40
PmR1; +; Gut	1348 ± 49	44	507 ± 31	25	38

Note: Data are means of final fresh weights and standard errors of means for *n* cultures incubated for 14 days at 25 ± 1°C.
[a]Secondary callus culture derived from heterozygous plant regenerated from mutant cell line selected for resistance to hydroxyurea.
Source: R. L. Keil and R. S. Chaleff (unpublished data).

considers that attempts have not yet been made to reproduce in vitro the spatial and geometric relationships in which cells normally develop in the plant. Thus variant phenotypes manifested by cultured cells and regenerated plants may very well reflect the responses of these cells, tissues, and plants to a foreign clime, just as the normal phenotype reflects development under normal conditions. Phenotypic abnormalities that are physiological responses to an altered environment are ephemeral and will persist only in that environment, disappearing when normal conditions are restored.

The responses of cultured tobacco tissues to the relative and absolute concentrations of auxin and cytokinin incorporated in the nutrient medium offer wonderful examples of physiological variation. The hormone composition of the medium determines the morphology, degree of friability, and organogenetic capability of the tobacco callus. For example, in the presence of moderate amounts of auxin growth of friable undifferentiated callus is favored by lower cytokinin concentrations whereas compact callus is produced by higher levels of cytokinin. Higher cytokinin levels also promote shoot formation while higher auxin concentrations stimulate differentiation of roots (Murashige and Skoog, 1962; Linsmaier and Skoog, 1965). All of these effects are fully reversible and are altered simply by adjustment of the hormone concentrations. (In contrast to the capability of callus to produce shoots and roots, the actual process of organogenesis represents an epigenetic change, since the differentiated structures continue to develop following transfer of the culture to a medium lacking hormones.)

The more deeply we pursue our analysis of the picloram-resistant mutants of tobacco, the more diverse become the aspects of cell culture with which we seem to become entangled. Looking again to these mutants, we find another variant phenotype that has a physiological basis. Among progeny of a heterozygous mutant plant (*PmR1*/+) regenerated from culture, tricotyledonous seedlings appeared at a higher frequency than they did among progeny of normal plants (Table 3.3). The tricotyly of the picloram-resistant mutants is not merely the result of passage through culture, since a normal picloram-sensitive plant regenerated from culture produced the same low frequency of tricots as did a normal plant grown from seed (Chaleff and Keil, unpublished results). Tricotyly occurs in many plant species. And as the cause of this phenomenon may be either genetic or physiological (DeVries, 1910; Haskell, 1954), several crosses were performed to determine its origin in this particular case. Since tricots occurred at approximately the same frequency among sensitive and resistant progeny of a heterozygous plant (*PmR1*/ +), tricotyly cannot be a pleiotropic effect of the *PmR1* allele or the result of mutation at a closely linked locus. Furthermore, self-fertilization of resistant and sensitive R_1 isolates (obtained from selfing the regenerated plant, see p. 93) produced the same frequency of tricot R_2 seedlings. Nor was a higher frequency found among progeny of two tricot R_1 seedlings that had been grown to maturity and selfed. In fact, it is clear that the frequency of tricots is influenced more by the generation than by the genotype of the parent plant. The frequency of approximately 20 percent that appears in the R_1 decreases to about 5% in the R_2 generation (R. S. Chaleff and R. L. Keil, unpublished results). From these observations, tricotyly in the picloram-resistant plants may be explained as a physiological response to picloram residues in the plant tissues. Picloram is an auxin analogue that is

Table 3.3. *Numbers and frequencies of tricotyledonous seedlings produced in crosses and in the R_1 and R_2 generations of picloram-resistant tobacco plants*

Cross	Picloram-resistant seedlings				Picloram-sensitive seedlings				Total
	Tricot	(%)	Dicot	Total	Tricot	(%)	Dicot	Total	Total
+/+ (from seed) selfed	0		0	0	0	(0)	669	669	669
Crosses with plants regenerated from culture									
+/+ (from callus) selfed	0		0	0	1	(0.2)	403	404	404
PmR1/+ selfed	120	(21)	449	569	31	(17)	153	184	753
+/+ × *PmR1*/+	0	(0)	244	244	0	(0)	241	241	485
R_1 isolates selfed									
PmR1/*PmR1* (isolate 4) selfed	5	(3)	152	157	0		0	0	157
PmR1/*PmR1* (isolate 10) selfed	7	(4)	190	197	0		0	0	197
PmR1/*PmR1* (isolate 11) selfed	7	(1)	689	696	0		0	0	696
PmR1/*PmR1* (isolate 12) selfed	14	(8)	167	181	0		0	0	181
+/+ (isolate 18) selfed	0		0	0	9	(5)	183	192	192
Tricot A selfed	36	(5)	714	750	3	(1)	236	239	989
Tricot B selfed	16	(6)	272	288	8	(5)	145	153	441

Source: R. S. Chaleff and R. L. Keil (unpublished data).

effective at extremely low concentrations (Kefford and Caso, 1966; Collins, Vian, and Phillips, 1978). It is likely that the herbicide accumulated by mutant cells during their selection and propagation in the presence of 500 μM picloram is responsible for the abnormal seedling morphology and that the frequency of tricots diminishes as the herbicide is diluted out during vegetative growth of the regenerated plant and in succeeding generations.

Merits and demerits of variability

From the documentation presented in this chapter it is clear that cultured plant cells are not the genetically, physiologically, and developmentally homogeneous population that appears typical for an analogous microbial system. Depending on one's intentions, the variability of cultured plant cells may prove a boon or a bane.

The first disadvantage resulting from the generation of variability in vitro is encountered in the analysis of a mutant phenotype. The expression of markers that are being used in a cross or protoplast fusion experiment can be modified by additional alterations arising in culture. Any such alterations can confuse interpretation of an experiment by making marker recognition difficult or ambiguous. Some traits are more prone to this problem than others. For example, coloration and leaf morphology vary greatly in regenerated plants and for this reason are unreliable markers. Modification of a mutant phenotype by additional variation also will complicate mutant selection experiments. Biochemical and genetic characterization will be more arduous if a new phenotype is the result of more than a single factor. Even if unexpected variability introduces changes that do not affect a mutant phenotype, one still has the task of resolving which of several metabolic and genetic changes actually are producing the altered phenotype. Until physiological modifications are eliminated and possible multiple genetic events are separated, no phenotypic alteration can be assigned a specific cause. For this reason, biochemical characterization of variant cell lines has not been very informative. In addition, plants regenerated from mutant cell lines must be outcrossed to place each mutation singly into an otherwise normal background before meaningful analyses can be performed. At such times one cannot help but wish that the experimental system possessed greater stability.

But yet other discomforts are caused by this unsought variability. Physiological, epigenetic, and genetic changes may interfere with normal metabolic and developmental processes. As a consequence, regenerated plants may be sterile, or callus cultures may lose the ability to differentiate altogether. It is frustrating to invest time and effort in selecting a variant cell line and then discover that only sterile plants or no plants at all can be regenerated. Furthermore, because of the inherent instability of cell cultures, it is not known whether the development of a plant or of fertile flowers is precluded by the selected alteration or by the occurrence of some other variation. One can only initiate a new cell line and begin the selection procedure again in the hope that the difficulties are not a pleiotropic effect of the desired mutation.

Progressive loss of totipotentiality of plant cells during a prolonged period of maintenance in culture is a common occurrence that still poses a severe and inscrutable obstacle to genetic studies. Loss of morphogenetic capacity has been correlated with karyotypic changes in callus cultures initiated from tobacco pith (Murashige and Nakano, 1967), pea root (Torrey, 1967), and carrot root (Smith and Street, 1974). Since tetraploid cells often prove capable of organogenesis, reduced morphogenetic capacity has been associated with higher ploidy levels (above $4n$), aneuploidy, and other chromosomal mutations that presumably occur during propagation in vitro. But the causes of totipotential decline are not solely genetic. In many cases, habituation, a phenomenon that is ostensibly epigenetic (see Chapter 2), is accompanied by loss of morphogenetic capacity (Gautheret, 1955). Morphogenetic capacity also is affected by the endogenous concentrations of metabolites and nutrients that may change during growth in culture, resulting in conditions unfavorable for organogenesis. For example, embryogenesis in cultured carrot cells is dependent on the quantity and form (oxidized vs. reduced) of nitrogen in the medium (Tazawa and Reinert, 1969). Thus, following a shift to a medium containing a high nitrogen concentration, embryogenetic competence was restored to long-term carrot cultures that apparently had lost this competence (Reinert, 1973). Of course, the evidence in all these studies is circumstantial, and at this point it seems impossible to establish a causal relationship between any single factor and a decline in the capacity for organogenesis or embryogenesis. Nevertheless, it is probable that genetic, epigenetic, and physiological changes occurring in culture all contribute to loss of morphogenetic capacity.

Spontaneous increases in chromosome numbers also can be rather maddening to someone who has taken the trouble to establish a haploid cell line for use in mutant selection experiments. Because a haploid cell contains only one copy of each gene, all mutations arising in that cell are expressed, since recessive alleles are not concealed by the presence of another gene copy. This advantage is lost when additional chromosomes are generated by endoreduplication or when rearrangements introduce duplications. The desirability of the haploid state for mutant isolation has inspired a search for a means to control the ploidy level of cultured cells. Gupta and Carlson (1972) suggested employing p-fluorophenylalanine

(PFP), a compound used to induce haploidization in the fungus *Aspergillus,* as an agent to select for growth of the haploid members of a heterogeneous population. If this method should prove effective, haploid cells would divide at a faster rate than cells of higher ploidies and would predominate in the culture, even though diploid and polyploid cells would continue to be produced by endoreduplication. It was reported that cell cultures initiated from pith explants of haploid *Nicotiana tabacum* plants proliferated on medium supplemented with PFP at 9 μg/ml, whereas growth of diploid cell cultures was inhibited completely. However, others were unable to repeat these results (Chaleff and Carlson, 1974; Zenk, 1974).

Experiments with cell cultures of *N. sylvestris* led Dix and Street (1974) to some interesting perceptions about the effects of PFP. Callus and suspension cultures derived from several haploid plants obtained by anther culture proved more resistant to growth inhibition by PFP than did diploid cultures. However, the various haploid cultures exhibited different degrees of resistance to the compound. Moreover, the ploidy composition of suspension cultures was unaffected by PFP. Now we already know that chromosome numbers are unstable in vitro and that cells of higher ploidies are generated rapidly. If the ploidy level does influence sensitivity to PFP, the distribution of karyotypes constituting such a heterogeneous suspension culture should shift toward lower ploidies during maintenance in a medium containing the analogue. But the diploid cells in the haploid-derived cell suspension continued to divide normally in the presence of a concentration of PFP that inhibited growth of cell cultures initiated from diploid explants. Thus, whereas cultures initiated from haploid plants are more resistant to PFP than are those initiated from diploid plants, haploid cells within a given cell line are no more resistant than isogenic diploid cells produced by chromosome doubling. On the basis of these observations, Dix and Street (1974) concluded that the greater resistance to PFP of the cell cultures derived from haploid plants is due to genotypic differences resulting from genetic segregation during meiosis and that the genotype of cultured cells rather than the ploidy level determines their sensitivity to growth inhibition by PFP. This interpretation received corroboration from experiments with cell suspension cultures of *Datura innoxia.* The proportions of haploid cells in three mixoploid populations were not altered by PFP, although these cell lines displayed differences in sensitivity to the compound (Evans and Gamborg, 1979).

However, the issue whether or not PFP favors growth of haploid cells is not yet resolved. In contrast to the results obtained with *N. sylvestris* and *D. innoxia,* Matthews and Vasil (1976) reported that PFP did not differentially affect growth of cell lines derived from haploid and diploid

N. tabacum plants, but it did influence the ploidy compositions of the cell lines. Cultures derived from a haploid plant and maintained for 6 weeks in the absence of PFP were composed of 3 percent haploid, 21 percent diploid, and 51 percent tetraploid cells, whereas cultures maintained for the same period in the presence of PFP at 10 μg/ml contained 56 percent haploid, 29 percent diploid, and 9 percent tetraploid cells.

And so the controversy over the utility of PFP still rages. In addition to the possibility that PFP stabilizes ploidy levels by promoting preferential growth of haploid cells, there is a suggestion that PFP might induce reductions in chromosome numbers. Niizeki (1976) reported the regeneration of a haploid plant of *Oryza punctata* from diploid callus that had been cultured in the presence of PFP. Although it was not shown that haploidization actually was effected by PFP, we are offered the exciting possibility of an alternative to anther culture as a method for obtaining haploid plants and as a means of producing haploid cells that will facilitate the development of parasexual genetic systems (see Chapter 5).

The occurrence of variability in vitro also has positive aspects. Spontaneous chromosome doubling in haploid cell cultures will produce homozygous diploid cells from which fertile plants can be regenerated (Nitsch, 1972). Doubling of chromosome numbers in vitro could also be employed to create fertile polyploid plants from sterile diploid progeny of interspecific crosses. Since in many species polyploidy is associated with improved yield, an additional agronomic application may be realized from the ability to recover polyploid plants from cultured cells. It is also conceivable that the karyotypic instability of cultured cells could be exploited to produce series of monosomic and trisomic plants and individuals possessing chromosome rearrangements that would be useful in genetic analyses. But by far the greatest importance of genetic variability in culture lies in its production of new mutant types. The apparently mutagenic effects of cell culture have provided a wealth of material for consideration in the next chapter.

4

Variants and mutants

By far the greatest experimental advantage of cell culture for the geneticist is that it makes large numbers of plant cells available for screening for variant types. Furthermore, growth of these cells under conditions that are better defined and more nearly homogeneous than can be achieved with whole plants facilitates direct selection and identification of variants. Plant mutants were always available to the geneticist, but these variant forms were identified on a visual basis rather than a biochemical basis. Mutants could be recognized and their inheritance studied because of their altered appearance, but in general they could not be selected deliberately as modifications of specific biochemical functions, and therefore in most cases the molecular basis of the change was neither understood nor accessible to study. For this reason, plants contributed primarily to the early development of the science of genetics, a period during which researchers were concerned with investigating the mechanisms of inheritance and the gross organization of the genome. But now the methods of cell culture can be applied to isolate biochemically defined mutants of plants that will permit us to probe into the molecular organization of these forms.

It is clear from the examples and arguments presented in the preceding two chapters that an altered phenotype can have a nongenetic basis. But because in many cases plants can be regenerated from cultured cells and a genetic analysis can be performed, a functional criterion by which to distinguish mutants and variants can be established. Only when a genetic change has been confirmed by analysis of inheritance through sexual crosses will a cell line or a plant be designated a mutant. In those cases in which it has been demonstrated that an altered phenotype has a physiological or epigenetic basis, or in which genetic analysis was not performed and the nature of the modification remains undetermined, an altered cell line is referred to as a variant. Because plant regeneration from cell or tissue cultures often cannot be accomplished, and as it is a rather slow process even when it can be, of the many variant cell lines that have been isolated thus far, only small numbers have been shown to be mutants. These mutants are listed in Table 4.1.

Table 4.1 *Mutant plants isolated by selection from cultured cells*

Mutant category	Mutant phenotype	Species	Mode of inheritance	References
Disease resistance	Resistance to methionine sulfoximine	*Nicotiana tabacum*	Two cases of a semi-dominant nuclear allele and one case of two recessive nuclear alleles	Carlson (1973a)
	Resistance to *Helminthosporium maydis* race T toxin	*Zea mays*	Maternal	Gengenbach, Green, and Donovan (1977)
Herbicide tolerance	Resistance to picloram	*N. tabacum*	Single dominant and semidominant nuclear alleles	Chaleff and Parsons (1978b)
Fungicide tolerance	Resistance to carboxin	*N. tabacum*	Nuclear dominant; number of alleles unknown[a]	Polacco and Polacco (1977), Polacco (pers. comm.)
Antibiotic resistance	Resistance to streptomycin	*N. tabacum* *N. tabacum* *Nicotiana sylvestris*	Maternal Maternal[b] Unknown[b]	Maliga et al. (1975) Umiel (1979) Maliga et al. (1980)
	Resistance to chloramphenicol	*N. sylvestris*	Unknown[a]	Maliga et al. (1980)
Purine and pyrimidine metabolism	Resistance to bromodeoxyuridine	*N. tabacum*	Nuclear dominant[a]	Márton and Maliga (1975)
	Resistance to hydroxyurea	*N. tabacum*	Single dominant nuclear allele	Keil and Chaleff (unpub. results)

Amino acid metabolism	Resistance to valine	*N. tabacum*	Single dominant and semidominant nuclear alleles	Bourgin (1978)
	Resistance to isonicotinic acid hydrazide	*N. tabacum*	Unknown[a]	Berlyn (pers. comm.)
	Resistance to glycine hydroxymate	*N. tabacum*	Unknown[a]	Lawyer, Berlyn, and Zelitch (pers. comm.)
	Resistance to lysine plus threonine	*Zea mays*	Single semidominant nuclear allele	Hibberd and Green (pers. comm.)
Carbon metabolism	Glycerol utilization	*N. tabacum*	Single dominant nuclear allele	Chaleff and Parsons (1978a)
Salt tolerance	Resistance to NaCl	*N. tabacum*	Unknown[b]	Nabors et al. (1980)
Auxotrophs	Biotin requirement	*N. tabacum*	Single recessive nuclear allele	Carlson (1970, pers. comm.)
	p-aminobenzoic acid requirement	*N. tabacum*	Single recessive nuclear allele	
	Arginine requirement	*N. tabacum*	Single recessive nuclear allele	
	Hypoxanthine requirement	*N. tabacum*	Two recessive nuclear alleles	

[a] Segregation has been observed among the progeny produced by some crosses, but this segregation is not in accordance with a conventional pattern and is insufficient to establish the mode of inheritance.

[b] The altered phenotype is transmitted to progeny, but critical crosses producing segregating populations have not yet been performed.

Choice of experimental material

The nature of the desired mutants will dictate the ploidy of the material to be used in a mutant isolation experiment. If one begins with a diploid cell line, recessive mutations will be visible only in cells possessing mutant alleles of both gene copies or in appropriate aneuploids. In the case of true diploids, mutant cells will appear at a frequency that is the square of the frequency of mutation of that particular locus. Therefore, when one has reason to believe that the desired mutation is recessive, or when one seeks to recover all possible types of mutations that can effect a given phenotype, haploid cells should be used. Haploid cell cultures can be obtained directly from the microspores contained within cultured anthers or from haploid plants produced either by anther culture or by any of several more traditional methods (parthenogenesis, gynogenesis, interspecific crosses) (Kasha, 1972). When it is anticipated that the desired mutant is dominant or semidominant, mutant recoveries may be equally efficient from diploid cells and haploid cells. The use of a diploid cell culture spares one the additional operation of doubling the chromosome number to produce a fertile plant. Diploid cells will be preferred to haploid cells if the desired mutation either is a recessive lethal or is dominant and is rarer than unwanted recessive mutations that cannot be discriminated by the selection procedure. For example, let us consider using an amino acid analogue that inhibits the activity of a biosynthetic enzyme to select mutant cells producing a feedback-insensitive form of the enzyme. It is probable that a very specific mutation is required to eliminate the feedback sensitivity of the enzyme without impairing its catalytic activity. In contrast, any number of mutations in the permease gene can destroy the activity of that protein and confer resistance to the analogue by preventing its assimilation. But as mutations of the permease structural gene are in general recessive, they will not be detected in a diploid cell containing a nonmutant allele of that gene. However, a diploid cell heterozygous at the locus encoding the allosterically regulated biosynthetic enzyme will produce the analogue-sensitive and -insensitive forms of the enzyme in equal quantities. And, as it is likely that a cell in which half of the activity of the biosynthetic enzyme is insensitive to inhibition will be able to grow in the presence of the analogue, the mutant allele will be dominant. In this case, therefore, the use of diploid cells might ensure the recovery of mutations in the structural gene of the biosynthetic enzyme rather than mutations in the structural gene of the permease.

Because of the prevalence of polyploidy throughout the plant kingdom, the production of true haploids containing only one basic chromosome set (monohaploids) is often difficult. In the case of an autotetraploid, such as potato or alfalfa, reduction to the gametic chromosome number creates a dihaploid that still possesses two similar sets of chromosomes. Dihaploids produced from an allotetraploid (e.g., *Nicotiana tabacum*) contain the genomes of two ancestral parents. If, during subsequent evolution, duplicate genes within the tetraploid genome diverge sufficiently, the species will become functionally diploid. But if the diploidization process is incomplete, the dihaploid may contain two copies of genes that differ slightly, but whose products still can perform the same function. In this circumstance, recessive mutations occurring in one gene copy will be concealed by the remaining normal allele. Analysis of a monosomic series of *N. tabacum* has revealed that certain recessive traits of this species are determined by single genes, whereas several others involve duplicate genes (Clausen and Cameron, 1944). Thus the question arises whether or not it is possible to detect recessive mutations in dihaploid individuals of an allotetraploid species such as *N. tabacum*. Doubtless the answer will depend on the species and the trait being considered. Carlson (1970) suggested that the recovery of only leaky auxotrophic mutants from selection experiments that he performed with cultured cells of a dihaploid *N. tabacum* plant resulted from the presence in the tobacco genome of duplicate copies of metabolically essential genes. Yet Müller and Grafe (1978), also working with dihaploid *N. tabacum* cultures, isolated recessive nitrate reductase–deficient cell lines that had no detectable nitrate reductase activity and were unable to utilize nitrate. These results are in agreement with the earlier findings of Clausen and Cameron (1944) indicating that the *N. tabacum* genome is partially but incompletely diploidized. Thus there is some justification for employing a true diploid species in developing a model genetic system in which the isolation of recessive mutations is important.

The terminology of ploidy can become rather complicated, and since the extent to which polyploid genomes have evolved toward diploidy is unknown, it is not necessarily helpful. Therefore, in the ensuing discussion, the term "haploid" will denote individuals containing the gametic chromosome number, and "diploid" will refer to individuals possessing the somatic chromosome number that is normal for the species.

Survey of variants and mutants isolated

The choices of variant types sought among cultured plant cells are influenced by the directions and achievements of microbial and animal cell genetics and by the central importance of plants to the welfare of man. Hence, resistance to antibiotics is pursued because the mechanisms of action of these compounds have been well characterized in bacteria. But in contrast to genetic studies of organisms used only for basic studies, many of the phenotypic modifications sought in plant cell cultures reflect man's constant striving to improve the nutritional and agronomic characteristics of these species. Because such variant forms are usually of lesser interest in microorganisms, often there is no relevant experience from microbial genetics on which to call. Once again we find our task more difficult because of that very individuality of plants that motivates our efforts.

Pigments and secondary metabolites

Because of their visually striking phenotype, mutants altered in pigment biosynthesis can be identified readily at the level of the intact plant, and therefore they were among the earliest plant mutant types to be isolated and characterized. Studies of such mutants inaugurated the biochemical genetic investigation of higher plants (Lawrence and Price, 1940; Haldane, 1941). Apparently the history of in vitro studies followed a similar course, since some of the first variant cell lines isolated were those differing from the normal in color. Altered coloration can be caused by either increased or decreased synthesis of any of several plant pigments, including chlorophyll. Such variant cell lines were isolated because, as in intact plants, they can be detected simply by visual inspection and also because of their possible relevance to efforts to amplify the production by cell cultures of secondary metabolites of commercial importance. Even if the particular pigment being overproduced by a variant cell line is of no value, study of the cell line may reveal methods for increasing the rate of synthesis of other products that are.

Perhaps the first report of a pigmented variant cell line concerned the isolation by Eichenberger (1951) of an orange-colored cell line of carrot containing a high level of carotene. Subsequently several carrot cell lines producing abnormal quantities of carotenoids were selected. One orange cell line obtained by Naef and Turian (1963) accumulates xanthophyll as well as α- and β-carotene. Other variant cell lines containing large amounts of lycopene have been characterized (Sugano, Miya, and Nishi, 1971; Nishi et al., 1974). Enormous variation in color and pigment content has been found among cell lines derived from a single carrot root. However, plants regenerated from these phenotypically dissimilar cell lines resemble the parent plant from which the cell cultures were obtained (Mok, Gabelman, and Skoog, 1976). Thus there is as yet no evidence that these altered patterns of pigment synthesis have a genetic basis.

Variation has also been observed in the anthocyanin content of cultured cells. Both anthocyanin-pigmented and colorless cells have been present in callus cultures derived from maize endosperm tissue. By selection of one or the other cell type, pigmented and nonpigmented cell lines have been established (Sternheimer, 1954). In addition, red-pigmented cell lines containing an anthocyanin different from that produced by the normal purple callus have been isolated from maize endosperm cultures. Colorless sectors have also appeared at low frequency in pigmented maize callus (Straus, 1958), and nonpigmented cell lines have been selected from anthocyanin-producing callus cultures of *Chenopodium amaranticolor* (Limbourg and Prevost, 1971). But selection of pigmented cell lines from nonpigmented callus cultures is more usual. Thus, anthocyanin-containing cell lines have been isolated from normally colorless cell cultures of *Haplopappus gracilis* (Blakely and Steward, 1961; Eriksson, 1967; Stickland and Sunderland, 1972), *Beta vulgaris* (Constabel, 1967), and *Daucus carota* (Alfermann and Reinhard, 1971). Clones accumulating large amounts of shikonins, which are red pigments of medicinal value, have been isolated from callus of *Lithospermum erythrorhizon*. Shikonins constituted as much as 4.8 percent of the dry weight of one such clone (Tabata et al., 1978).

Selection for more highly pigmented cell lines can also be used to increase the levels of other compounds whose synthesis is correlated with production of the pigment. Thus, realizing that in heterogeneous callus cultures of *Macleaya microcarpa* alkaloids are produced by small scattered populations of cells that are colored by a nonalkaloid yellow pigment, Koblitz and associates (1975), by visually selecting the yellow color, isolated cell lines containing high alkaloid concentrations.

The green color of chlorophyll also invites selection for altered production of this pigment. Green cell lines have been isolated from nonpigmented callus tissue of *Taxus* (Tulecke, 1959), tobacco (Venketeswaran, 1965), and *Atropa belladonna* (Davey, Fowler, and Street, 1971). And as in the case of anthocyanin, selection can be practiced in the opposite direction by isolating albinos among populations of pigmented cells or regenerated plants. By such procedures albino cell lines have been recovered from a mutagenized population of *Datura innoxia* protoplasts (Krumbiegel, 1979), and albino plants have been regenerated from cultured anthers of *N. tabacum* (Nitsch, 1972) and from protoplasts of *D. innoxia* (Schieder, 1976).

Variant cell lines that accumulate greater quantities of colorless secondary metabolites for which direct selection procedures are not yet available can be isolated by assaying large numbers of clones for their content of the desired compound. In this manner, Zenk and associates (1977) employed a radioimmunoassay to screen clones of *Catharanthus roseus* and identify cell lines possessing two alkaloids (ajmalicine and serpentine) in a combined amount equal to 1.3% of the dry weight. In addition, cell lines of *N. tabacum* producing high levels of nicotine have been isolated by assaying the nicotine content of more than 1000 clones (Tabata et al., 1978). At present, the design of selection procedures is limited by insufficient knowledge of plant metabolism and its regulation.

Hence, mass screening often is the only available means of identifying a variant cell line that overproduces a particular compound. But as more information about plant metabolism is acquired, it will become possible to devise schemes for selecting increasing numbers of variant types.

Habituation

Other characteristics of cultured plant cells in which variation can be discerned fairly readily are their requirements for nutritional and hormonal supplements. Gautheret (1946) first observed that callus cultures of *Scorzonera,* which normally require an exogenous supply of auxin, would occasionally (after several passages in culture) acquire the ability to grow in the absence of this hormone. Originally named *accoutumance à l'hétéro-auxin* by Gautheret (1946), this phenomenon is now referred to as habituation (see Chapter 2). Auxin-habituated cell lines have also been isolated from callus cultures of *Vitis vinifera* (Morel, 1947), tobacco (Fox, 1963), *Acer pseudoplatanus* (Lescure and Péaud-Lenoël, 1967), *Crepis capillaris* (Sacristán and Wendt-Gallitelli, 1971), and many other species (Gautheret, 1955). Tobacco cell cultures can also lose their requirement for cytokinin (Fox, 1963; DeMarsac and Jouanneau, 1972; Binns and Meins, 1973). As discussed in Chapter 2, present evidence suggests that habituation is caused by epigenetic changes affecting the rates of hormone synthesis and/or degradation.

More recently it has been discovered that variant cell lines of carrot and potato selected for resistance to 5-methyltryptophan may be auxin-autotrophic. An exogenous supply of auxin is not required for growth of one potato cell line and growth of 5 of 10 carrot cell lines that contain both a feedback-insensitive form of anthranilate synthetase and an elevated endogenous concentration of free tryptophan. Interestingly, none of several tryptophan-overproducing cell lines of tobacco has proved auxin-autotrophic, even though tryptophan or indole can relieve the auxin requirement of cultured tobacco cells (Widholm, 1977a). Because tryptophan and indole are precursors of indole-3 acetic acid (IAA), it is not surprising that increasing their endogenous rate of production or providing them in the medium will eliminate the requirement for an exogenous auxin supply by stimulating biosynthesis of this hormone. It has been shown that an auxin-autotrophic carrot cell line that had been selected for resistance to 5-methyltryptophan does, in fact, contain an increased amount of IAA (Sung, 1979). However, it is not understood why only some of the carrot cell lines and none of the tobacco cell lines that accumulate tryptophan are auxin-autotrophic. Unfortunately, because excess auxin prevents plant regeneration (and hence genetic analysis), it is not known whether or not the altered phenotypes are caused by mutation (and, if so, the number of mutations involved).

Nutritional variants have been isolated that have developed the capacity for growth on medium lacking a vitamin supplement that formerly had been essential. Gautheret (1950) was unable to initiate cultures of *Salix caprea* (goat willow) in the absence of pantothenic acid, but after these cultures were established, their maintenance no longer required this vitamin. Similarly, cultures obtained from tumorous tissues of *Rumex acetosa* initially expressed a thiamine requirement that was lost after several years of propagation in vitro (Nickell, 1961). More recently, a thiamine-independent soybean cell line was isolated by first shifting callus to a medium supplemented with one thiamine precursor and then transferring three surviving cell lines to a medium containing a second precursor (Ikeda, Ojima, and Ohira, 1979). In all three reports the authors compared the loss of a vitamin requirement with habituation. Although it is quite possible that elevated rates of vitamin synthesis resulted from epigenetic changes, evidence regarding the mechanism of the observed modifications is not yet available.

Resistance to antibiotics

Because the mechanisms of action of most antibiotics are known from work with microbial systems, these compounds were obvious choices for use in selecting variant plant cell lines. Having isolated a variant plant cell line resistant to an antibiotic, one enjoys the important advantage of knowing generally among which cellular components the altered function might be identified.

Cell lines resistant to streptomycin were first isolated from haploid callus cultures of *Petunia hybrida* (Binding, Binding, and Straub, 1970; Binding, 1972). Shortly thereafter, streptomycin-resistant cell lines of *N. tabacum* and *N. sylvestris* were obtained by Maliga, Sz. Breznovits, and Márton (1973). In these experiments, callus initiated from haploid plants was placed on medium containing a hormone composition that normally induces shoot formation and a concentration of streptomycin (0.5 mg/ml) that inhibits cell division, organogenesis, and chlorophyll synthesis. Green tissue sectors from which shoots were developing appeared after 6 to 8 weeks. Diploid plants were regenerated from one resistant *N. tabacum* cell line. These plants were backcrossed reciprocally, and progeny from these crosses were again crossed reciprocally with a normal plant and selfed. Crosses were scored by evaluating the ability of seedlings to form callus on medium containing streptomycin. The responses of progeny to streptomycin were always the same as that of the female parent (Maliga et al., 1975). This consistently maternal pattern of inheritance established streptomycin resistance as a genetically stable extranuclear mutation. That this mutation is located in the chloroplast genome was revealed by comparison of ribosomal proteins from chloro-

plasts of normal plants and mutant plants by two-dimensional gel electrophoresis. A difference was detected in the mobility of 1 of the 67
proteins resolved. These experiments also demonstrated that in tobacco,
as in bacteria, streptomycin inhibits protein synthesis by binding a protein
component of the ribosome and that resistance is achieved by mutational
alteration of that protein (Yurina, Odintsova, and Maliga, 1978).

Streptomycin-resistant mutants have also been selected from diploid
callus cultures of *N. tabacum* (Umiel and Goldner, 1976). In this case the
phenotype of progeny of regenerated plants was determined by germinating seeds in an aqueous streptomycin solution. Yellow seedlings were
obtained from seeds from normal plants, whereas only green seedlings
were obtained from seeds produced by crosses in which a plant regenerated from a streptomycin-resistant cell line was the female parent.
Because the mutant cell line had been isolated from cell cultures derived
from a male sterile plant, crosses using regenerated mutant plants as the
male parent could not be made. However, backcross progeny produced
by fertilization of the regenerated plant with nonmutant pollen also
produced only mutant offspring when pollinated by a normal male.
Therefore this streptomycin-resistant mutation likewise is inherited maternally (Umiel, 1979).

Ultrastructural studies of the two streptomycin-resistant *N. tabacum*
mutants revealed what appears to be an interesting difference between
them. Streptomycin causes degeneration of both chloroplasts and
mitochondria of normal tobacco callus (Zamski and Umiel, 1978). Plastids
in the streptomycin-resistant callus tissue isolated by Umiel and Goldner
(1976) were not severely affected by the drug; unlike those in normal
tissue, they developed thylakoid membranes and grana in its presence.
However, mitochondria of both tissues appeared to be equally sensitive to
streptomycin (Zamski and Umiel, 1978). In contrast, Maliga and associates (1975) reported (electron micrographic evidence was not presented)
that neither chloroplasts nor mitochondria in leaves of their mutant plants
were damaged by streptomycin. It remains to be demonstrated that, in
this latter tobacco mutant, resistance of both organelles results from but
one mutational event. A single chloroplast mutation that renders both
mitochondrial and chloroplast protein synthesis insensitive to inhibition
by streptomycin has been identified in *Chlamydomonas reinhardtii*. On
the basis of this observation, it was proposed that the chloroplast genome
of *C. reinhardtii* encodes a streptomycin-binding protein that is a component of both chloroplast and mitochondrial ribosomes (Conde et al.,
1975).

Quite a different result was obtained by selecting for streptomycin
resistance among cell cultures derived from a haploid *N. sylvestris* plant.
Fertile plants regenerated from variant cell lines produced resistant
progeny when self-fertilized, but not when crossed with a normal plant.

These observations suggest, but are insufficient to establish, that streptomycin resistance of these *N. sylvestris* cell lines is determined by a
recessive nuclear mutation (Maliga et al., 1980).

Streptomycin resistance was also discovered somewhat unexpectedly
in a haploid cell line of *N. sylvestris* that had been selected for resistance
to kanamycin, another antibiotic that interferes with the functioning of
70S ribosomes. But because plants could not be regenerated from these
cell lines, it is not known if their resistance to the two antibiotics is due to
more than a single event (Dix, Joó, and Maliga, 1977). Kanamycin-
resistant cell lines of *N. tabacum* have also been isolated. Secondary
callus cultures initiated from plants regenerated from some, but not all, of
the variant cell lines exhibited resistance to the antibiotic (Maliga et al.,
1980).

Cycloheximide is an antibiotic that inhibits protein synthesis on cytoplasmic eukaryotic ribosomes. Cell lines resistant to cycloheximide have
been isolated from cultures obtained from haploid *N. tabacum* protoplasts, but the resistance phenotype was maintained only as long as cell
lines were propagated in the presence of the drug (Maliga et al., 1976).
These experiments have been described in a different context in Chapter
2. Cycloheximide-resistant cell lines of carrot have also been selected.
The results of one study were similar to those obtained with tobacco in
that the resistance phenotype of the selected cell lines was unstable and
disappeared in the absence of the drug. When plants were regenerated
from these cell lines on medium supplemented with cycloheximide,
secondary callus cultures initiated from these plants also proved resistant.
However, no genetic studies were performed (Gresshoff, 1979). In
another study, resistance of variant carrot cell lines to cycloheximide was
reported to be stable and to be expressed even following long periods of
culture on a medium lacking the antibiotic (Sung, 1976). The basis of these
variations in the stability of cycloheximide resistance remains to be
explored.

Maliga and associates (1980) recently reported the isolation of cell lines
resistant to chloramphenicol from initially haploid *N. sylvestris* cell
cultures. A few of the plants regenerated from one variant cell line gave
rise to chloramphenicol-resistant secondary callus cultures. Of 135 progeny obtained from crosses with these regenerated plants, only two
seedlings proved resistant to chloramphenicol. Obviously further studies
are required to reveal the basis of this phenotype.

Purine and pyrimidine metabolism

In mammalian cell culture systems, cell lines lacking certain enzymes of
purine and pyrimidine metabolism have been isolated by selecting for

resistance to analogues of these compounds. These deficiencies have been exploited widely as genetic markers in cell fusion and gene transfer experiments (as reviewed by Chu and Powell, 1976). The successes realized with mammalian cell lines resistant to nucleic acid base analogues have motivated efforts to obtain similar variants of cultured plant cells.

Cell lines resistant to the thymidine analogue 5-bromodeoxyuridine (BUdR) have been isolated from cultured soybean (Ohyama, 1974, 1976) and sycamore (Bright and Northcote, 1974) cells and from callus cultures initiated from haploid (Maliga, Márton, and Sz. Breznovits, 1973) and diploid (Lescure, 1973) *N. tabacum* plants. Although the molecular basis of resistance of the variant sycamore cell line was never determined, it is unlikely that it is inability to assimilate the analogue, since both the thymidine kinase activity and its capacity to take up thymidine from the medium were normal (Bright and Northcote, 1974). Both of these functions were also intact in the two soybean cell lines that were selected for resistance to BUdR, although in one of these cell lines (BU-5) the amount of ATP-independent thymidine phosphorylating activity was appreciably reduced. But as BUdR was incorporated freely into DNA of BU-5 cells, the resistance of this cell line clearly was not due to exclusion of the analogue (Ohyama, 1974). In contrast, the second variant soybean cell line (BU-54), which was selected from BU-5 for resistance to a higher concentration of the analogue, incorporated BUdR into DNA only in the presence of fluorodeoxyuridine (FUdR). This cell line was also resistant to FUdR, an inhibitor of thymidylate synthetase, and to aminopterin, an inhibitor of dihydrofolate reductase. Resistance to these inhibitors probably was due to the greatly elevated levels of these enzymes found in BU-54 cells. But another consequence of possessing a superabundance of these enzymes was an increased capacity to synthesize thymidylate. On this basis it was postulated that the BU-54 cells were protected from BUdR by overproducing thymidylate, which competitively excluded the analogue from incorporation into DNA. This postulate is consistent with the observation that BUdR is incorporated into DNA of BU-54 cells in the presence of FUdR when thymidylate synthesis is inhibited, but not in its absence (Ohyama, 1976).

Resistance to BUdR in a variant tobacco cell line is achieved by a quite different mechanism. These BUdR-resistant cells have normal thymidine uptake and phosphorylation capacities and, unlike the BU-54 soybean cells, are sensitive to FUdR. Since the toxicity of FUdR is reversed by thymidine, excess amounts of thymidylate apparently are not synthesized by these cells and therefore are not the basis on which resistance to BUdR is effected. Cells of this variant tobacco culture incorporate BUdR into DNA to the extent that approximately 15% of the thymidine residues are displaced by the analogue (Márton et al., 1978).

Plants have been regenerated from several of the BUdR-resistant tobacco cell lines that were isolated from a haploid-derived callus culture. However, only plants produced from one cell line flowered, and because all of these plants were triploid, their fertility was low. Growth tests of secondary callus cultures established from these plants revealed that not all were BUdR-resistant, and therefore the original variant cell line must have been chimeral. The resistant plants were crossed, and the phenotypes of progeny seedlings were assigned by initiating callus cultures and determining their growth responses to BUdR. Self-fertilization produced only BUdR-resistant progeny, whereas both resistant and sensitive progeny were obtained from reciprocal backcrosses. One would like to think that BUdR resistance is due to a single dominant nuclear mutation that occurred after polyploidization of the cell culture and that therefore the regenerated plants are heterozygous. But if this were the case, self-fertilization would be expected to produce some sensitive progeny. Since the variant cell culture produced sensitive plants and therefore contained normal cells, another possible explanation of the recovery of sensitive progeny from the backcrosses is that the resistant plants regenerated were chimeral and that in some cases germ lines arose from nonmutant cells (Márton and Maliga, 1975).

The failure of selection for resistance to BUdR to yield a thymidine kinase–deficient plant mutant has disappointed many. But there is really no justification for despair. The biochemical basis of resistance has been explored in only two soybean cell lines (one derivative of the other) (Ohyama, 1974, 1976), one sycamore cell line (Bright and Northcote, 1974), and two tobacco cell lines (Márton et al., 1978). This sample represents a rather small number when one considers the many other mechanisms of BUdR resistance that have been demonstrated in mammalian cell cultures, including those that do not prevent incorporation of BUdR into DNA (Hsu and Somers, 1962; Davidson and Bick, 1973; Bick and Davidson, 1974). Clearly, many more variant plant cell lines must be isolated before the full range of possibilities has been explored.

Knowledge that resistance of mammalian cells to the purine analogues 8-azaguanine and 6-thioguanine results from loss of hypoxanthine-guanine phosphoribosyltransferase (HGPRT) activity (Brockman and Anderson, 1963; Sharp, Capecchi, and Capecchi, 1973) encouraged selection of plant cell lines resistant to these compounds. Azaguanine-resistant cell lines of *N. tabacum* (Lescure, 1973), sycamore (Bright and Northcote, 1975), *H. gracilis* (Horsch and Jones, 1978), and soybean (Weber and Lark, 1979) and one thioguanine-resistant soybean cell line (Weber and Lark, 1979) have been isolated. The variant sycamore cells possess approximately half the normal level of HGPRT activity, which is what would be expected if one of two structural genes contained within a diploid genome had mutated (Bright and Northcote, 1975). HGPRT activity is reduced approximately 30% in the resistant *Haplopappus* cell line (Horsch and Jones, 1978). But one should hesitate before presuming that these changes in HGPRT activity are responsible for resistance of the cells to azaguanine. Because of the enormous variability that occurs in cultured plant cells (see Chapter 3), fluctuations must occur in the levels of many enzymes of such cells. Measurement of the activity of a single enzyme is insufficient to establish a change in that activity as the basis of an altered cellular phenotype. At least the levels of other metabolically related enzymes should be shown to have remained unaltered. Of course, more definitive evidence of association of analogue resistance with a change in the activity of a particular enzyme would be provided by regenerating plants and demonstrating that in sexual crosses the two characteristics cosegregate. One can benefit from knowledge acquired from experiments with cultured mammalian cells without accepting the disadvantages of those systems. Certainly, established plant cell lines or cultures of species possessing small chromosome numbers have advantages for particular types of studies, but for studies involving the isolation of mutants, the advantages of genetic analysis are of such great importance that cell cultures capable of giving rise to plants should be used in preference to cultures from which plants cannot be regenerated.

Subsequent selection experiments with the azaguanine-resistant *Haplopappus* cell line yielded a cell line that was also resistant to 6-azauracil (Jones and Hann, 1979). Resistance to another pyrimidine analogue, 5-fluorouracil, has been selected among cultured carrot cells (Sung, 1976), and cell lines of *D. innoxia* resistant to aminopterin have been isolated (Mastrangelo and Smith, 1977). This compound is a folic acid antagonist that inhibits the production of tetrahydrofolate, which is required in thymidylate and purine biosynthesis. Neither biochemical nor genetic studies have yet been performed with any of these variants, although plants have been regenerated from the aminopterin-resistant *Datura* cell lines.

Once again we turn for inspiration to the array of variants that have been isolated from cultured mammalian cells. And we find that by means of selection for resistance to hydroxyurea, Chinese hamster ovary cell lines have been isolated that synthesize either an altered form of ribonucleotide reductase (Lewis and Wright, 1974) or increased amounts of this enzyme (Lewis and Wright, 1979). Plating diploid cells of *N. tabacum* on a medium supplemented with 100 μM hydroxyurea yielded 15 resistant cell lines. Secondary callus cultures initiated from plants regenerated from 10 of these variant cell lines expressed resistance to hydroxyurea. These regenerated plants first were crossed with normal plants before

Table 4.2. *Progeny obtained from crosses with heterozygous plants* (HuR/+) *derived from two hydroxyurea-resistant cell lines*

	No. of resistant individuals		No. of sensitive individuals	
	Observed	Expected	Observed	Expected
HuR1/+ selfed	39	45	20	15
+/+ × HuR1/+	15	16	17	16
HuR1/+ × +/+	19	16	13	16
HuR9/+ selfed	53	48	11	16
+/+ × HuR9/+	14	16	18	16
HuR9/+ × +/+	15	14.5	14	14.5

Source: R. L. Keil and R. S. Chaleff (unpublished data).

genetic analysis was performed. Table 4.2 presents the results of crosses with the heterozygotes derived from plants regenerated from two hydroxyurea-resistant cell lines. Because the hydroxyurea-resistance phenotype is not expressed by plants and therefore must be determined from the growth responses of derivative callus cultures, the numbers of progeny scored are necessarily small. Nevertheless, it is evident that in both cases resistance is conferred by a single dominant nuclear allele (R. L. Keil and R. S. Chaleff, unpublished results).

Disease resistance

The selection of plant mutants exhibiting greater resistance to disease is our first consideration of an application of cell culture that does not seek merely to reproduce experiments performed with microbial and mammalian cells but that addresses problems and properties peculiar to the plant kingdom. The screening by conventional procedures of populations of plants that had been regenerated from cultured cells or protoplasts for susceptibility to disease was discussed in Chapter 3. It has been reported that sugarcane plants resistant to eyespot (Heinz, 1973) and Fiji (Krishnamurthi and Tlaskal, 1974) diseases and potato plants resistant to *Alternaria solani* (Matern, Strobel, and Shepard, 1978) have been recovered by this procedure. But another approach to the isolation of varieties relatively disease-tolerant is by selecting directly among cultured cells for resistance to the toxin elaborated by the pathogenic organism. In this method cells are plated on a medium containing a lethal concentration of the toxin, and rare growing colonies are selected. Although this in vitro

selection procedure could be much more efficient than screening populations of whole plants, it has several disadvantages that limit its applicability. First, resistance exhibited by the cultured cell may not be expressed
by the mature plant (see Chapter 2). The second constraint is far more
severe: selection for disease resistance in vitro can be performed only in
those cases in which damage to the plant results from the action of a toxin
that is produced by the pathogen. If the disease symptoms are caused
primarily by a toxin, that toxin should display the same specificity as the
pathogen in eliciting those symptoms. Unfortunately, only a few such
host-specific toxins have yet been characterized (Scheffer, 1976).

In the first selection of disease resistance among cultured plant cells,
Carlson (1973a) recovered mutants of *N. tabacum* resistant to wildfire
disease. This disease of tobacco is caused by a bacterial pathogen,
Pseudomonas tabaci, which produces a toxin structurally similar to
methionine (Stewart, 1971). Populations of mutagenized haploid protoplasts and cells first were cultured in a nonselective medium, and after 2
weeks this initial medium was overlaid with a second medium containing
an inhibitory concentration of methionine sulfoximine (MSO). MSO, an
analogue of the wildfire toxin, elicits formation of the same characteristic
chlorotic halos on tobacco leaves as does the natural bacterial toxin
(Braun, 1955). Resistant calluses were selected, and diploid plants were
regenerated from three cell lines that continued to express resistance
stably on further testing. Leaves on these plants were inoculated with *P.
tabaci* and a solution of MSO to determine their sensitivities to both
agents. The chlorosis that normally develops on leaves of the parent plant
in response to these substances did not appear on leaves of plants regenerated from the three MSO-resistant cell lines (Figure 4.1). However,
small necrotic spots that did develop at the point of inoculation with *P.
tabaci* were similar to those obtained from infection of tobacco with *P.
angulata* (Carlson, 1973a). As *P. angulata* is a variety of *P. tabaci* that
does not produce toxin (Braun, 1937), it is apparent that selection for
resistance to MSO yielded plants that were insensitive to the action of the

FIGURE 4.1. Response to infection by *Pseudomonas tabaci* and to application of
methionine sulfoximine (MSO) of leaves from normal and mutant *Nicotiana
tabacum* plants. *a:* Control leaf from a normal *N. tabacum* cv. Havana Wisconsin
38 plant. *b:* Leaf from mutant 1. *c:* Leaf from mutant 2. *d:* Leaf from mutant 3.
e: Leaf from cv. Burley 21, which is naturally resistant to *P. tabaci.* Leaves were
inoculated at two points in the apical region with a culture of *P. tabaci* in nutrient
broth. An aliquot (0.1 ml) of a 100 μM solution of MSO was applied to the left side
of the basal portion of each leaf. Uninoculated nutrient broth was applied to the
right basal region. Reactions of leaves were scored after 6 days. (From Carlson,
1973a. Copyright 1973 by The American Association for the Advancement of
Science.)

toxin itself but that still were susceptible to other deleterious effects of bacterial infection.

Heterozygous plants were produced by crossing regenerated homozygous diploid plants with a normal plant. The heterozygous F_1 plants were selfed, and segregation was observed among the progeny by plating seeds on a medium supplemented with 10 mM MSO. The heterozygote formed from mutant 1 produced sensitive, intermediate, and resistant progeny at a ratio of approximately 9:6:1. This result suggests that resistance of mutant 1 was caused by two independently segregating recessive loci with additive effects. Leaves of homozygous mutant 1 plants contained normal amounts of all free amino acids. Self-fertilization of heterozygous plants formed from mutants 2 and 3 yielded sensitive, intermediate, and resistant seedlings at a ratio of 1:2:1, indicating that in these cases resistance was due to a single semidominant nuclear mutation. In homozygous mutant 2 and mutant 3 plants the endogenous concentrations of free methionine were five times higher than normal, whereas the levels of other free amino acids remained unchanged (Carlson, 1973a). The levels of free methionine in the heterozygous mutant 2 and mutant 3 plants were intermediate between those in the normal plant and those in the respective homozygous mutants (Chaleff and Carlson, 1975). The intermediate methionine levels of these heterozygotes are consistent with the observed transmission of the resistance phenotype as a single semidominant allele, and they suggest that the elevated methionine pools, if not in some way part of the mechanism of resistance, are at least a pleiotropic effect of the mutations conferring that resistance.

Because of the susceptibility of maize plants having Texas male-sterile cytoplasm (*cms*-T) to the fungal pathogen *Helminthosporium maydis* race T, it seemed appropriate to attempt selection of cell lines resistant to the *H. maydis* toxin from callus cultures derived from sensitive male-sterile plants. Such an experiment is made feasible by the specificity of *H. maydis* toxin, which inhibits growth of *cms*-T callus cultures but does not affect growth of callus from male-fertile plants resistant to the fungus (Gengenbach and Green, 1975). Cell lines resistant to a normally lethal toxin concentration were isolated from *cms*-T callus by initially culturing callus in the presence of a slightly inhibitory amount of toxin and then sequentially transferring vigorously growing callus pieces to media containing increasing concentrations of toxin. Plants were regenerated from these cell lines, and their phenotypes were determined from the responses of leaf tissues to direct application of a toxin solution. Of 65 plants regenerated from callus that had survived at least five cycles of selection, all responded as resistant to the toxin. But 52 of these toxin-resistant plants were male-fertile, and the sterility of the remaining 13 was due to something other than *cms*-T cytoplasm, since none of these plants could function as either a male or female parent, and in most cases floral organs were deformed. It is probable that aberrations occurring during propagation in vitro were responsible for the sterility of these individuals (see Chapter 3). In reciprocal crosses of fertile toxin-resistant regenerated plants with *cms*-T plants, both the fertility and the toxin response among the progeny always resembled those of the maternal parent. These progeny were also infected with *H. maydis* spores, and in all cases the reactions to the fungus and to the partially purified toxin solution were the same. Thus, in contrast to the results obtained by selection of MSO-resistant tobacco cell lines, selection for resistance of maize callus to the *H. maydis* toxin produced plants that were also resistant to the causal organism. Unfortunately, however, in these experiments male sterility proved inseparable from susceptibility to *H. maydis,* and therefore no practical benefit was realized. These results can be interpreted in two ways. One possibility is that both male sterility and toxin susceptibility are determined by a single extranuclear locus and that any mutation of this locus that confers toxin resistance also restores male fertility. Another explanation is that the cytoplasm of the parental *cms*-T plant from which the cell cultures were initiated was heterogeneous and contained both *cms*-T and toxin-resistant mitochondrial genomes. In this case, growth in the presence of the toxin would select cells in which the toxin-resistant mitochondria would predominate and from which the *cms*-T mitochondria would have been largely eliminated. Male fertility would also be restored as a consequence of this process if it were determined either by the same locus that confers toxin resistance or by another locus that does not recombine with that for toxin resistance (Gengenbach, Green, and Donovan, 1977).

Recently the sequences of small segments of the mitochondrial genomes of progeny of several regenerated toxin-resistant plants have been compared by digesting mitochondrial DNA purified from these plants with various restriction endonucleases. Each of these enzymes cleaves DNA at a specific and characteristic nucleotide sequence. If the mitochondrial genomes of all of these plants are identical, enzymatic digestion will produce the same fragments in each case, and hence resolution of the fragments of each genome by gel electrophoresis will produce the same pattern. However, if changes in sequence have occurred that have eliminated some cleavage sites or introduced others, new fragment patterns will appear. By this method, differences have been detected in the mitochondrial genomes of the progeny of regenerated plants and the *cms*-T line from which they were derived. Although distinct fragment patterns were found for the progeny of three independently derived resistant plants, digests of all these plants lacked one particular fragment that was present in digests of *cms*-T plants. This information still provides no clue to the origin of the genomic differences, but it is tempting to speculate that the endonuclease recognition site that produces the one fragment missing from progeny of all three regenerated plants analyzed is involved in some way in conditioning both male sterility and toxin sensitivity (B. G. Gengenbach and D. R. Pring, personal communication).

To date I am aware of only one other attempt to isolate disease-tolerant varieties from cultured plant cells. In these experiments, Behnke (1979) selected a small number of potato cell lines that survived transfer to a medium made toxic to normal callus by addition of the filtrate of a *Phytophthora infestans* culture. Plants were regenerated from many of these resistant cell lines, and the ability of leaves from these plants to form callus on selective medium was employed as a test of the stability of the resistance phenotype. After 6 weeks callus was produced by approximately 40 percent of the leaves of plants regenerated from resistant cell lines and by only 5 percent of control leaves. As yet, however, resistance of the regenerated plants to infection by *P. infestans* has not been determined, nor have genetic crosses been performed.

Nitrate assimilation

The experiments described thus far have not involved the application of more than a single selection scheme to a given pathway. One feels strongly the constraints imposed on the spectrum of variants that can be recovered by the small number of selection methods that are feasible at present. In most cases the types of variants isolated are dictated by the availability of a selective agent, rather than by the types of variants desired. But within the nitrate assimilation pathway are opportunities for selecting several kinds of variants in which both regulatory and structural functions have been modified. The ability to select alterations of several different steps of one pathway elevates genetic studies with plant cells to a new level of sophistication. Our experimental approach at last becomes directed, rather than desultory, and analysis of a biochemical pathway and its regulation can be attempted.

In higher plants, nitrate is taken up and converted to ammonium by the sequence of reactions depicted in Figure 4.2. After entering the cell via a permease, nitrate is reduced to nitrite by nitrate reductase. Nitrate reductase is a complex enzyme composed of several subunits that in addition to catalyzing the NAD(P)H-dependent reduction of nitrate, can utilize reduced viologen dyes, $FMNH_2$, or $FADH_2$ as electron donors for nitrate reduction and can mediate the reduction of cytochrome c by

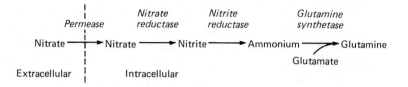

FIGURE 4.2. Pathway for assimilation of nitrate by a plant cell. The dashed vertical line represents the cell membrane. Ammonium is most probably assimilated through combination with glutamate to form glutamine (Miflin and Lea, 1976).

NAD(P)H* (Hewitt, Hucklesby, and Notton, 1976). Finally, nitrite is reduced to ammonium by nitrite reductase, which is located in the chloroplasts. This pathway is important agronomically because most plants assimilate nitrogen primarily in the form of nitrate. An understanding of the genetic organization of the nitrate pathway will provide a basis for constructing mutants that use fertilizer more efficiently.

When tobacco cells are cultured in a medium containing nitrate as the sole nitrogen source, certain single amino acids, such as threonine, and mixtures of amino acids inhibit growth. A variant cell line resistant to threonine-induced inhibition of growth on nitrate was isolated and characterized by Heimer and Filner (1970). Although casein hydrolysate inhibited both nitrate reductase synthesis and nitrate uptake in normal cells, in the variant cells only the development of nitrate reductase activity was inhibited, not nitrate uptake. From these observations it was concluded that regulation of nitrate uptake had been altered in the variant.

In experiments with bacteria (Puig and Azoulay, 1967; Stouthamer, 1969), fungi (Cove, 1976), and seeds of higher plants (Oostindiër-Braaksma and Feenstra, 1973; Tokarev and Shumnyi, 1977), chlorate has been used to select mutations in the nitrate assimilation pathway. In most cases it appears that chlorate itself is not toxic but is converted by nitrate reductase to the toxic chlorite. However, the mode of action of chlorate may be considerably more complex. Evidence suggests that in *Aspergillus* chlorate is toxic because it prevents catabolism of nitrogen compounds (Cove, 1976). In any case, mutants deficient in nitrate reductase are selected by plating cells on chlorate-supplemented medium.

By applying this selection method to haploid cell cultures of *N. tabacum,* Müller and Grafe (1978) isolated resistant cell lines lacking

*NADH, reduced form of nicotinamide adenine dinucleotide; NADPH, reduced form of nicotinamide adenine dinucleotide phosphate; FMNH$_2$, reduced form of flavin mononucleotide; FADH$_2$, reduced form of flavin adenine dinucleotide.

nitrate reductase activity. Variants having no detectable nitrate reductase activity could not utilize nitrate, but they could grow on medium containing reduced nitrogen compounds, such as amino acids. Among the many variants that were recovered, the degree of resistance to chlorate corresponded inversely to the residual amount of nitrate reductase activity. Several variants (*cnx*) lacked both xanthine dehydrogenase and nitrate reductase activities, indicating that these cell lines were probably deficient for a molybdenum-containing cofactor that is common to both enzymes. However, extracts of *cnx* cells contained normal levels of nitrate reductase–associated cytochrome c reductase activity. Other variants (*nia*) possessed xanthine dehydrogenase activity and lacked catalytic activities normally associated with nitrate reductase (Table 4.3) (Mendel and Müller, 1979). That independent functions were affected in the two classes of variants, as suggested by the biochemical assays, was confirmed by in vitro complementation studies. Low levels of nitrate reductase activity were restored by mixing extracts of the two types of variants (Mendel and Müller, 1978). Partial reconstitution of nitrate reductase activity was also achieved in vivo by fusing protoplasts obtained from *cnx* and *nia* cell lines. These somatic hybrids were also able to utilize nitrate and form plants, capabilities that were not expressed by the variants (Glimelius et al., 1978). Plants could also be regenerated from variant cell cultures containing residual amounts of nitrate reductase activity. The stable expression of the reduced nitrate reductase activities in these plants and in secondary callus cultures initiated from them causes one to suspect that the chlorate-resistant cell lines were true mutants (Müller and Grafe, 1978). Of course, proof of this supposition must await genetic analysis of the somatically produced hybrid plants. Presumably, the nitrate reductase–deficient variants that have been isolated could now be used in reconstruction experiments to develop other selection methods based on the inability of such variants to grow on nitrate as the sole source of nitrogen. Cell lines lacking permease or nitrite reductase activity conceivably might be isolated by this procedure. It should also be possible to select directly for revertants by plating mutant cells that are unable to assimilate nitrate on medium containing nitrate as the sole nitrogen source. Surely we can look forward to exciting developments in studies of the nitrate assimilation pathway in the years ahead.

The work of Müller and Grafe (1978) illustrates more clearly than any other the importance of isolating and characterizing a large number of variants in each system. When examining only a small number of variants, one cannot expect to obtain the full spectrum of different types that might be available from a selection procedure. In fact, it is an almost diabolical realization, long known to those working with microbes but not yet fully appreciated by those struggling with plants, that the particular type of

Table 4.3. *Nitrate reductase, nitrite reductase, and xanthine dehydrogenase activities (as units per mg protein) in normal and chlorate-resistant cell lines of* N. tabacum

Cell line	Induction[a]	Nitrate reductase			NADH-cytochrome c reductase	Nitrite reductase	Xanthine dehydrogenase
		NADH[b]	FADH$_2$[c]	BVH[d]			
Normal	−	102	59	31	20	142	+
	+	545	327	169	59	401	
nia-02	−	0	0	0	13	168	+
	+	0	0	0	9	428	
nia-04	−	0	0	0	16	108	+
	+	0	0	0	17	388	
nia-95	−	0	18	11	14	272	+
	+	0	46	25	21	627	
cnx-68/1	−	0	0	0	27	245	−
	+	0	0	0	101	283	
cnx-101	−	0	0	0	68	326	−
	+	0	0	0	82	836	
cnx-109	−	0	0	0	28	487	−
	+	0	0	0	64	528	

[a]Uninduced activities were assayed in extracts of cells that had been grown on medium containing an amino acid mixture as the sole nitrogen source. Induced activities were assayed in extracts prepared 5 hours after the addition of KNO_3 (final concentration 50 mM) to the medium.
[b]Nitrate reductase activity measured using NADH as electron donor.
[c]Nitrate reductase activity measured using FADH$_2$ as electron donor.
[d]Nitrate reductase activity measured using reduced benzyl viologen as electron donor.
Source: Mendel and Müller (1979).

variant or mutant being sought may be recovered at a lower frequency than other less interesting types. In such a case it is unlikely that the desired variant can be obtained without examining a large number of isolates.

Sulfate assimilation

At present, only a single publication has reported the use of variant cell lines to explore the sulfate assimilation pathway in higher plants. These variants were selected by transferring cells of a threonine-resistant tobacco cell line (see this chapter, "Nitrate assimilation") to a liquid medium containing both selenocystine and selenomethionine. Selecting resistance to both toxic selenium analogues should avoid recovery of variants that are blocked in the uptake of these compounds or in which regulation of the biosynthesis of the single amino acids is altered. Instead, this scheme favors the isolation of variants accumulating both cysteine and methionine. Overproduction of these two amino acids is expected as a consequence of the derepression of sulfate assimilation. Three variant cell lines were selected, but only one proved stably resistant to both selenocystine and selenomethionine. This cell line was also more sensitive to selenate, a toxic analogue of sulfate, during growth on a medium containing cysteine as the sulfur source. Since cysteine normally inhibits sulfate assimilation, sensitivity to selenate suggests that this pathway is no longer repressed in the variant. The two other cell lines were resistant to selenocystine and selenate, but not to selenomethionine. As yet, no direct measurements have been reported of either sulfate uptake rates or ATP sulfurylase activities of the variant cell lines (Flashman and Filner, 1978). Nevertheless, plants are being regenerated, and it is expected that analysis of these and similar variants will contribute much to our knowledge of sulfur metabolism in higher plants.

Amino acid metabolism

Because of their enormous importance in microbial and mammalian cell culture systems, it was to be expected that variants altered in some aspect of amino acid metabolism would be sought among cultured plant cells. In addition, such experiments with plant cells are motivated by the prospect that they will contribute either directly or indirectly, by yielding knowledge of metabolic regulatory mechanisms, to improvements in the nutritional values of food crops.

Resistance to analogues of amino acids is the simplest and most widely employed procedure for selecting mutations affecting the control of amino acid biosynthesis or catabolism. The toxicity of amino acid analogues to cultured plant cells is due to either their incorporation into proteins in lieu of the natural amino acid or their inhibition of allosterically regulated biosynthetic enzymes. Resistance can be conferred by mutations that introduce a means of degrading the analogue or excluding it from the cell. In cases in which an analogue interferes with cellular metabolism by binding an allosteric enzyme, inhibition can be relieved by a mutation that reduces the affinity of the enzyme for the analogue. In addition, this type of mutation, as well as any other that increases the rate of synthesis of the natural amino acid, will overcome toxicity caused by incorporation of an analogue into protein by diluting its effective intracellular concentration. Mutations that enable the protein synthetic machinery to discriminate against an analogue will also prevent its incorporation into protein.

The isolation by Carlson (1973a) of tobacco mutants resistant to wildfire disease, as discussed in a preceding section ("Disease resistance"), provided the first demonstration that genetically stable mutants of higher plants in which regulation of amino acid biosynthesis is altered can be selected directly in vitro. As in microbial systems, selection for resistance to a structural analogue (methionine sulfoximine) of an amino acid (methionine) led to the recovery of mutants containing elevated endogenous concentrations of the natural amino acid.

Even predating the methionine sulfoximine–resistant tobacco mutants was Widholm's (1972a,b) selection from presumably diploid tobacco and carrot cell cultures of variants resistant to 5-methyltryptophan. This analogue prevents growth by inhibiting the activity of anthranilate synthetase, the first enzyme specific to tryptophan biosynthesis (Figure 4.3), thereby blocking the production of tryptophan (Widholm, 1972c). Crude extracts of the analogue-resistant cell lines contained a species of anthranilate synthetase that was less sensitive than the normal enzyme to inhibition by tryptophan and 5-methyltryptophan. Endogenous concentrations of free tryptophan in variant cell lines of tobacco and carrot were 10-fold and 27-fold higher, respectively, than in the parental cell lines (Widholm, 1972a,b). Subsequently, variant tobacco cell lines having even higher (33-fold) levels of free tryptophan were reported (Widholm, 1977b). Initially, plants could be regenerated only from analogue-resistant carrot cell lines possessing a normal feedback-sensitive form of anthranilate synthetase, not from cell lines producing a feedback-insensitive form of the enzyme. Secondary cell cultures derived from the regenerated plants resembled the originally isolated variant cell line in displaying resistance to growth inhibition by 5-methyltryptophan and in possessing moderately elevated levels of free tryptophan (ninefold above normal), reduced capacity for tryptophan uptake, and normal anthranilate synthetase activity. It was suggested that resistance of these cell lines to growth inhibition by 5-methyltryptophan was due to diminished permeability to

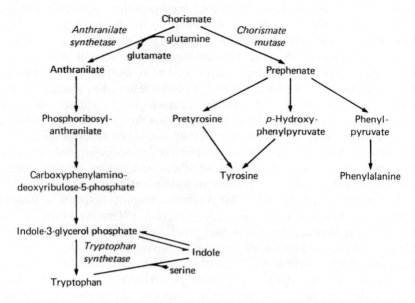

FIGURE 4.3. Pathway for biosynthesis of tryptophan, tyrosine, and phenyl-alanine.

the inhibitor (Widholm, 1974b). Whatever the mechanism effecting analogue resistance, the fact that this trait was also expressed in cell cultures derived from regenerated plants demonstrates that the variant phenotype possesses the stability expected of a mutation. Unfortunately, however, genetic crosses with the regenerated plants, which would have revealed the basis of the resistance phenotype, were not performed. More recently, plants were regenerated from 5-methyltryptophan-resistant tobacco cell lines containing a feedback-insensitive form of anthranilate synthetase. Although the altered enzyme was produced by secondary callus cultures initiated from the regenerated plant, it was not present in the leaves of this plant (Widholm, 1978a). This evidence is insufficient to determine whether analogue resistance results from mutational alteration of an isozyme of anthranilate synthetase that is produced only in callus tissue or from a regulatory phenomenon (genetic or epigenetic) that causes expression in callus of a feedback-insensitive isozyme of anthranilate synthetase that normally is not produced by such tissue.

Two isozymes of anthranilate synthetase, one sensitive and one resistant to inhibition by tryptophan, have been detected in cultured potato cells. Normal cell lines contain more of the feedback-sensitive form, whereas 5-methyltryptophan-resistant cell lines contain more of the feedback-insensitive form. The endogenous concentration of free tryptophan in one analogue-resistant potato cell line was 48-fold higher than normal (Carlson and Widholm, 1978). Presumably, resistance of these cell lines to 5-methyltryptophan is achieved by increasing the rate of synthesis of an anthranilate synthetase isozyme that is not inhibited by the analogue and correspondingly decreasing the rate of synthesis of the sensitive isozyme, but the molecular basis of this change is not known.

This same kind of experiment has been performed yet again with cultured cells of *Arabidopsis thaliana* (Negrutiu, Jacobs, and Cattoir, 1978) and *Catharanthus roseus* (Scott, Mizukami, and Lee, 1979). In some 5-methyltryptophan-resistant *Arabidopsis* cell lines the capacity to assimilate the analogue is apparently reduced, whereas the endogenous concentration of free tryptophan is elevated in others (Negrutiu, Jacobs, and Cattoir, 1978). *Catharanthus* cell lines resistant to 5-methyltryptophan possess a feedback-insensitive form of anthranilate synthetase and therefore also accumulate excess tryptophan. Interestingly, the resistant *Catharanthus* cells contain both 100 percent more tryptophan synthetase activity and 50 percent more anthranilate synthetase activity than normal cells (Scott, Mizukami, and Lee, 1979). But, once again, the failure to regenerate plants leaves us without knowledge of the number or types of events responsible for these manifold changes.

Cell lines resistant to *p*-fluorophenylalanine (PFP), an analogue of phenylalanine, have been selected from cell cultures of tobacco, carrot (Palmer and Widholm, 1975), sycamore (Gathercole and Street, 1976, 1978), and *D. innoxia* (Evans and Gamborg, 1979). PFP-resistant carrot cell lines contain elevated endogenous concentrations of free phenylalanine (sixfold) and tyrosine (threefold) and a chorismate mutase (Figure 4.3) activity that, like the normal carrot enzyme, is not inhibited by phenylalanine or tyrosine. A reduced rate of uptake of the analogue is at least in part responsible for resistance of the carrot cell line to the analogue. In resistant tobacco cells, the level of tyrosine (threefold), but not that of phenylalanine, is increased. Although uptake of PFP by the variant tobacco cell line is decreased, its analogue resistance can be attributed to a higher specific activity of chorismate mutase and to a form of the enzyme that differs from the activity present in normal cells in being insensitive to inhibition by phenylalanine, tyrosine, or PFP (Palmer and Widholm, 1975). The normal intracellular concentration of phenylalanine in the resistant cells is explained by the presence of higher levels of phenylalanine ammonia lyase activity, which prevent accumulation of the amino acid by converting it into phenolic compounds (Berlin and Widholm, 1977). One of the PFP-resistant sycamore cell lines contains normal amounts of free phenylalanine and tyrosine and increased phenylalanine ammonia lyase activity. Resistance of this cell line could be the result of either reduced uptake or increased catabolism of the

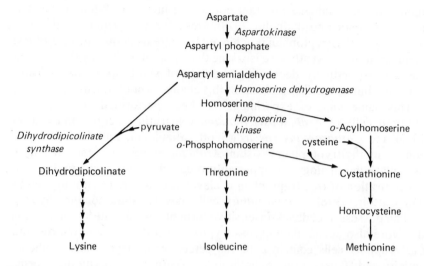

FIGURE 4.4. Pathway for biosynthesis of lysine, methionine, threonine, and isoleucine in higher plants.

analogue (Gathercole and Street, 1976). Higher levels of both free phenylalanine and tyrosine are present in a second variant sycamore cell line (Gathercole and Street, 1978). Although the PFP-resistant *D. innoxia* cell line has not been characterized biochemically, shoots have been regenerated, and further studies are to be anticipated (Evans and Gamborg, 1979).

The series of reactions by which lysine, threonine, isoleucine, and methionine are synthesized from aspartate (Figure 4.4) is of particular interest, since all four of these amino acids are essential to human nutrition. Moreover, lysine generally is the nutritionally limiting amino acid in cereals, whereas small amounts of methionine limit the quality of legume seed proteins. The hope, then, is to select regulatory mutations that will cause increased production of these amino acids. The biosynthesis of lysine and methionine is regulated by complex mechanisms that differ from one species to another. This complexity is compounded by the production during plant development of several enzyme forms that exhibit different sensitivities to inhibition by the end products of the pathway (see Chapter 2) (Bryan, 1976). To make its point simply, this discussion will consider only those control mechanisms that have been described for the few cereal species studied thus far (maize, wheat, and barley).

The three enzymes of lysine and methionine synthesis that apparently function as key control points are aspartokinase, dihydrodipicolinate synthase, and homoserine dehydrogenase (Figure 4.4). Maize seems to possess only a single form of aspartokinase (Cheshire and Miflin, 1975; Bryan, 1976), which, like the activities extracted from barley and wheat, is inhibited by lysine alone (Bryan et al., 1970; Cheshire and Miflin, 1975; Aarnes, 1977; Shewry and Miflin, 1977; Gengenbach et al., 1978; Bright, Shewry, and Miflin, 1978). Three forms of homoserine dehydrogenase differing in their sensitivities to inhibition by threonine have been identified in maize. The proportion of one form that is most strongly inhibited by threonine decreases during development, resulting in a loss of sensitivity to threonine during seedling growth (DiCamelli and Bryan, 1975; Matthews, Gurman, and Bryan, 1975). Two forms of homoserine dehydrogenase possessing different sensitivities to threonine have also been extracted from barley, although no study has been made of the relative amounts of the two forms present at various stages of development (Aarnes, 1977). As dihydrodipicolinate synthetase is the first enzyme specific to the lysine branch of the pathway, it is not surprising that the maize enzyme is inhibited solely by lysine (Cheshire and Miflin, 1975).

Because of these allosteric control mechanisms, the flow of carbon through the aspartate pathway can be interrupted by exposing cells to either a lysine analogue (which presumably inhibits aspartokinase and dihydrodipicolinate synthase) or an excess of lysine plus threonine. Some dicotyledonous plants possess an aspartokinase that is inhibited by both lysine and threonine (Dunham and Bryan, 1969, 1971; Wong and Dennis, 1973). But since the aspartokinase activity found in cereals is not affected by threonine, it is speculated that in these species lysine and threonine act singly to inhibit aspartokinase and homoserine dehydrogenase, respectively (Bright, Wood, and Miflin, 1978). In both cases the observed inhibition of cell growth results from the cessation of methionine synthesis, since this inhibition is reversible by the addition of methionine (Dunham and Bryan, 1969; Wong and Dennis, 1973; Henke et al., 1974; Green and Phillips, 1974; Chaleff and Carlson, 1975; Bright, Wood, and Miflin, 1978).

Selection of mutations affecting regulation of the aspartate pathway was first attempted by plating mutagenized rice cell suspension cultures on medium containing the lysine analogue *S*-2-aminoethylcysteine (SAEC). Three SAEC-resistant cell lines that contained elevated levels of free lysine, methionine, isoleucine, leucine, valine, and several other amino acids were isolated. The total (free plus incorporated) amounts of lysine, isoleucine, leucine, and valine were also increased. Growth of one variant cell line was not inhibited by a concentration of lysine plus threonine that completely inhibited growth of the parental cell line (Chaleff and Carlson, 1975). But as plants could not be regenerated from the variant cell lines, it was never determined if these alterations were due to a mutational event or if they could influence the amino acid composition of the seed.

Cell lines of *N. tabacum* (Widholm, 1976) and *A. thaliana* (Negrutiu, Jacobs, and Cattoir, 1978) resistant to SAEC have also been selected. Growth of a *N. tabacum* cell line selected for resistance to δ-hydroxylysine, another lysine analogue, was not inhibited by lysine plus threonine. The amount of free lysine accumulated by both analogue-resistant *N. tabacum* cell lines was more than 10-fold higher than normal (Widholm, 1976). Resistance of the variant *Arabidopsis* cell lines to SAEC ostensibly was due to a different mechanism. During growth on SAEC, more of the analogue was accumulated (but less was incorporated into protein) by the variant than by the normal cells, and in contrast to other SAEC-resistant cell lines that have been reported, the *Arabidopsis* variants contained reduced amounts of free lysine (Negrutiu, Jacobs, and Cattoir, 1978).

The inhibition of cell growth by lysine plus threonine and the reversal of this inhibition by methionine suggested that this toxic mixture of amino acids could be used to select methionine-overproducing variants (Green and Phillips, 1974). That lysine plus threonine might prove more effective than SAEC as an agent for selecting such variants was indicated by the observation that the growth differential between a variant rice cell line selected for resistance to SAEC and the parental cell line was larger on lysine plus threonine than on the analogue (Chaleff and Carlson, 1975).

Recently, a variant cell line of maize resistant to lysine plus threonine was isolated from callus cultures initiated from scutellar tissue. This variant cell line possessed an aspartokinase activity that was less sensitive than the normal activity to inhibition by lysine. In addition, cells resistant to lysine plus threonine contained increased amounts of free threonine, methionine, lysine, and isoleucine. Secondary callus cultures initiated from shoots regenerated from the variant callus cultures were also resistant to lysine plus threonine, but unfortunately none of these shoots proved fertile (Hibberd et al., 1980). Fertile plants were regenerated from a second variant maize cell line resistant to lysine plus threonine. Reciprocal backcrosses of these regenerated plants, which presumably were diploid and heterozygous, yielded both resistant and sensitive progeny. A resistant plant obtained from a backcross produced a majority of resistant progeny when selfed and nearly equal numbers of resistant and sensitive progeny when crossed again with a normal plant (Table 4.4). These results, in combination with the biochemical data, clearly show that resistance to lysine plus threonine is due to a single semidominant nuclear mutation. Homozygous mutant kernels contained nearly 80-fold more free threonine, threefold more free serine, and fourfold more free methionine than did normal kernels. In heterozygous kernels the amount of free

Table 4.4. *Segregation of resistance to lysine plus threonine among progeny of a heterozygous* LT19/ + *maize plant*

	No. of resistant individuals		No. of sensitive individuals	
Cross	Observed	Expected	Observed	Expected
LT19/ + selfed	71	83.25	40	27.75
+/ + × *LT19/* +	47	51.5	56	51.5

Source: K. A. Hibberd and C. E. Green (personal communication).

threonine was increased 30-fold (K. A. Hibberd and C. E. Green, personal communication). The discovery of a mutation in maize conferring resistance to lysine plus threonine that affects regulation of the biosynthesis of the aspartate-derived amino acids and the amino acid composition of the seed is an exciting one indeed. The dream of employing cell culture to achieve genetic improvements in crop plants is fast becoming a reality.

Selection for resistance to lysine plus threonine is currently being pursued as well with haploid rice callus derived by anther culture. Plantlets have been regenerated directly on selective medium (Figure 4.5) and now are being grown to maturity for further studies (R. S. Chaleff, unpublished results).

Another approach to the isolation of methionine-overproducing plant cell lines is to use resistance to methionine analogues to select variants in which regulation of the methionine-specific branch of the pathway is altered. Thus the free methionine content has been elevated more than 10-fold in an ethionine-resistant carrot cell line (Widholm, 1976). J. Madison and J. Thompson (personal communication) have identified 6 soybean cell lines that accumulate excess free methionine from among more than 100 cell lines that had been selected for resistance to ethionine. Whereas normal cells contain 6 to 12 nmoles of methionine per gram of fresh weight, cells of the 6 variant cultures contain 40 to 80 nmoles of methionine per gram of fresh weight. Abnormally high levels of S-adenosylmethionine are also present in most of these variants.

Although the synthesis of valine and leucine and that of isoleucine proceed through two different series of intermediates, a common set of enzymes catalyzes the first three reactions of both biosynthetic sequences (Figure 4.6). Hence these sequences are regarded as a single branched pathway. In higher plants the first enzyme of this pathway, acetohydroxy acid synthase, is inhibited by leucine and valine (Miflin and Cave, 1972). Therefore, as is the case for the branched pathway by which the

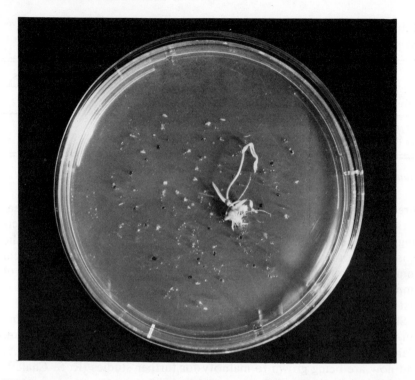

FIGURE 4.5. Rice plantlet regenerating from callus directly on selective medium. Callus obtained by anther culture was transferred to medium supplemented with 1 mM lysine plus 0.5 mM threonine. Most of the callus died, but occasionally a resistant cell line appeared and gave rise to a plantlet. (R. S. Chaleff, unpublished results.)

aspartate-derived amino acids are produced, an excess of one end product of the branched pathway will block synthesis of the amino acids produced by the other branches. Specifically, it has been shown that valine inhibits division of tobacco protoplasts and that this inhibition is reversed by isoleucine (Bourgin, 1976). Mutants of *N. tabacum* resistant to growth inhibition by valine have been selected by Bourgin (1978). In these experiments protoplasts obtained from a haploid plant formed by anther culture were mutagenized by ultraviolet irradiation. After 1 month callus was transferred to medium containing a toxic concentration of valine, and resistant calluses were isolated as they appeared. Adult plants were regenerated from two valine-resistant cell lines. Fertile diploid plants were obtained from one cell line *(Val^r-2)*, but another cell line *(Val^r-1)*, yielded only male-sterile plants with between 41 and 49 chromosomes. Only fully resistant progeny were obtained by selfing *Val^r-2* plants, and all

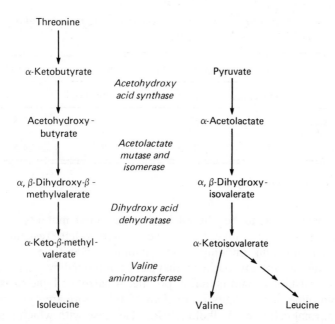

FIGURE 4.6. Pathway for biosynthesis of isoleucine, leucine, and valine.

progeny produced from reciprocal backcrosses with normal plants displayed a lower degree of resistance. These results suggest, but are insufficient to prove, that the valine resistance of the Val^r-2 plants is due to a single semidominant nuclear allele. Fully resistant progeny resulted from pollination of a Val^r-1 individual by a normal plant. Self-fertilization of one of these F_1 plants produced 34 resistant and 16 sensitive progeny (in close agreement with a 3:1 ratio), and reciprocal crosses with a normal plant yielded nearly equal numbers of resistant and sensitive seedlings (Table 4.5). Thus, in the case of Val^r-1 plants, resistance is conferred by a single dominant nuclear allele. Because biochemical studies have not been reported at this time, nothing is known about the mechanism(s) of valine resistance. As discussed earlier, resistance to growth inhibition by valine could result from synthesis of a feedback-insensitive form of acetohydroxy acid synthase or from degradation or exclusion of the amino acid. But another possibility is suggested by the ability of valine to interfere with nitrate assimilation when nitrate is the sole nitrogen source and the ability of isoleucine to antagonize this effect of valine (Filner, 1966; Behrend and Mateles, 1975). If in Bourgin's experiments valine were inhibiting cell growth by this mechanism, one would expect regulation of nitrate assimilation, rather than regulation of leucine-isoleucine-

Table 4.5. *Segregation for valine resistance among progeny of a heterozygous* Val^r-1/ + *tobacco plant*

Cross	No. of resistant individuals		No. of sensitive individuals	
	Observed	Expected	Observed	Expected
Val^r-1/ + selfed	34	37.5	16	12.5
Val^r-1/ + × +/ +	21	22	23	22
+/ + × Val^r-1/ +	27	24.5	22	24.5

Source: Bourgin (1978).

valine biosynthesis, to be altered in the recovered mutants. This last explanation appears unlikely, however, since valine resistance was selected on a medium containing both ammonium and nitrate.

One other biosynthetic pathway that has been probed by selection of an analogue-resistant variant is that of proline. A carrot cell line resistant to hydroxyproline accumulates more than 10-fold more free proline than does a normal cell line (Widholm, 1976). The ease with which plant cell lines resistant to amino acid analogues can be obtained has been demonstrated by the sequential selection of a carrot cell line resistant to *p*-fluorophenylalanine, ethionine, *S*-2-aminoethylcysteine, and 5-methyltryptophan (Widholm, 1978b). Indeed, it is evident from the brief summary presented in this chapter that the genetic and biochemical characterization of such variants is lagging considerably behind their isolation.

The prospect of increasing photosynthetic efficiency by decreasing the rate of oxidation of photosynthate has attracted attention to the photorespiratory pathway. In the first steps of this catabolic pathway, ribulose-1,5-bisphosphate is converted into glycine, from which serine and CO_2 are then formed. This last reaction is inhibited by isonicotinic acid hydrazide (INH) and glycine hydroxamate (GH), both of which thereby cause the accumulation of glycine (Asada et al., 1965; Bird et al., 1972; Lawyer and Zelitch, 1979). As these compounds also inhibit the growth of cultured tobacco cells, cell lines resistant to this growth inhibition have been selected in the hope that such variants will provide insight into the functioning of the photorespiratory pathway.

Cell lines resistant to INH were selected by plating suspension cultures derived from haploid *N. tabacum* plants on medium supplemented with the inhibitor. Of 58 cell lines isolated, 20 continued to display resistance to INH following a period of growth in its absence. Vigorous plants were regenerated from 10 variant cell lines, and secondary callus cultures initiated from these plants were tested for INH resistance. Resistant callus was obtained from all plants regenerated from several cell lines and from at least some plants regenerated from the other cell lines. Many of these plants apparently had diploidized spontaneously and were able to produce seed. Callus cultures initiated from progeny seedlings were scored for growth on INH. Resistant progeny were obtained from the self-fertilization of plants regenerated from five variant cell lines (M. B. Berlyn, personal communication). Although additional crosses will be required for definitive proof, the preliminary results are certainly in accord with a genetic basis for INH resistance.

Seven tobacco variants stably resistant to GH were also isolated from cell cultures initiated from haploid plants. None of these cell lines exhibited an abnormal rate of uptake or degradation of GH, nor did any of the variants contain an altered form of glycine decarboxylase, the enzyme inhibited by GH. However, GH-resistant cell lines did accumulate higher levels of all free amino acids and take up glycine and alanine from the medium at a greater rate than did normal cells. Another characteristic of all but one GH-resistant cell line was that, unlike normal cells, their growth was not inhibited by an excess of glycine or alanine. Plants were regenerated from five of the seven variant cell lines. Although tissues of these plants did not contain increased concentrations of free amino acids, growth of secondary callus cultures and of calluses initiated from progeny of several regenerated plants was resistant to GH. In contrast to the examples considered earlier, in which cell lines resistant to amino acid analogues were found to possess increased amounts of the corresponding natural amino acids, GH-resistant cell lines contained higher levels of all of the free amino acids that were analyzed. But if an increase in glycine alone is toxic, or if GH also functions as an analogue of amino acids other than glycine, such a general increase in the intracellular concentrations of free amino acids might be the simplest mechanism by which to effect resistance to the analogue (Lawyer, Berlyn, and Zelitch, 1980).

Carbon metabolism

Perhaps the first indication of the possibility of recovering plant cell lines altered in their ability to utilize various carbon sources came from the study of clones derived from cultured grape tissue. Arya, Hildebrandt, and Riker (1962) discovered that whereas cells of one clone grew best on sucrose or lactose, other cloned cell lines grew better on glucose, mannose, or galactose than on sucrose or lactose. Cells of yet another clone responded equally well to glucose and lactose and grew only slowly on sucrose. Subsequently, similar observations were made with cloned cell lines of the tobacco hybrid *N. tabacum* × *N. glutinosa* (Sievert and Hildebrandt, 1965).

In several cases variant cell lines have been selected for their capacity to grow on medium containing a carbon source that cannot be utilized by normal cells. The first such experiment produced tobacco cell lines able to utilize lactose as the sole carbon source. These variant cell lines, which were isolated by transferring calluses to medium containing a mixture of 2 percent lactose and 2 percent sucrose, possessed elevated levels (fourfold to eightfold above normal) of β-galactosidase activity (Kapitsa, Kulinich, and Vinetskii, 1977). And because soybean cell suspension cultures normally grow very slowly in maltose medium, it has been possible to recover variants that grow rapidly in this medium (Limberg, Cress, and Lark, 1979).

Maretzki and Thom (1978) have isolated a sugarcane cell line able to grow on galactose as the sole carbon source. This trait is stable in that following propagation in a sucrose medium for a period of 6 weeks the cells still retain their capacity for growth on galactose. The probable basis of the ability of the variant cells to utilize galactose was revealed by a comparison of the activities of the enzymes of the galactose-utilization pathway present in variant cells growing on galactose and in normal cells that had been either maintained on sucrose or transferred to (but obviously not grown on) galactose. Whereas galactokinase activity was only slightly higher in cells of the variant culture, these cells contained 10-fold more uridine diphosphate (UDP) galactose-4-epimerase than did normal cells incubated in either sucrose or galactose medium. Increased activity of UDP-galactose-4-epimerase, which catalyzes the conversion of UDP-galactose to UDP-glucose, prevented the accumulation of UDP-galactose that occurs in normal cells transferred to galactose medium and enhanced the capacity of the variant cells to channel this metabolite into glycolysis.

A *N. tabacum* cell line capable of growth on glycerol has also been isolated. Because this cell line (designated *Gut*) responded normally to several other sugars that were tested, it appeared that the modification specifically affected glycerol assimilation. It was believed that if the *Gut* cell line proved to be a true mutant, it would be heterozygous, since it had been isolated from a diploid culture. Accordingly, approximately equal numbers of *Gut* and normal progeny were obtained when plants regenerated from the *Gut* cell line were backcrossed reciprocally with normal plants. However, self-fertilization of these regenerated plants produced 11 *Gut* and 9 normal progeny. Although the discrepancy between this result and the theoretical pattern of a 3:1 ratio between *Gut* and normal expected for segregation of a dominant nuclear allele is not statistically

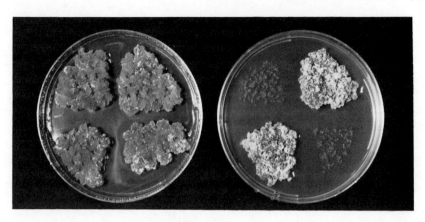

FIGURE 4.7. Segregation of the *Gut* phenotype in a cross of a heterozygous *(Gut/ +)* with a normal (+/ +) plant. Callus cultures were initiated from progeny seedlings and transferred to media containing either sucrose (left) or glycerol (right) as sole carbon source. Each quadrant of a plate contains callus derived from a separate seedling. Corresponding quadrants of the two plates were inoculated with callus pieces of the same origin. (R. S. Chaleff, unpublished results.)

significant, these numbers are in closer agreement with a 2:1 ratio (Chaleff and Parsons, 1978a). Such data could be due simply to the small sample size, which is an inevitable consequence of a scoring procedure that requires that callus be initiated from individual progeny seedlings and then tested for growth on glycerol (Figure 4.7). However, another possibility is that zygotes or seeds homozygous for the *Gut* mutation are inviable. The results of the reciprocal backcrosses with a heterozygous *Gut/ +* mutant indicate that the *Gut* mutation segregates normally in meiosis and does not interfere with the function or development of the gametophytes, where it is present in a hemizygous state. Culturing anthers of heterozygous plants produced 36 *Gut* and 31 + haploid individuals (Table 3.1), confirming that hemizygous *Gut* plants are capable of normal growth and development and that the *Gut* phenotype is caused by mutation of only a single locus. (If two mutations had been required to produce the *Gut* phenotype, it would have been expressed by only one-fourth of the haploids.) Another selfing was attempted, but this time a plant doubly heterozygous for picloram resistance (p. 85) and glycerol utilization was used, and larger numbers of progeny were scored. Although the ratio between picloram-resistant and picloram-sensitive progeny was 3:1, indicating normal meiotic segregation, the recovery of 28 *Gut* and 22 normal individuals from this cross was a significant

Table 4.6. *Progeny obtained from self-fertilization of a doubly heterozygous* PmR1/+; Gut/+ *plant*

	Number of progeny of phenotype			
	PmR;Gut	*PmR;+*	*+;Gut*	*+;+*
Observed results	25	16	3	6
Expected results if homozygotes are viable	28	9.5	9.5	3
Expected results if homozygous *Gut/Gut* individuals are inviable	25	13	8	4

Source: R. S. Chaleff (unpublished data).

deviation from a 3:1 ratio (Table 4.6). At this point one is left with the intriguing possibility that the *Gut* mutation has a dual character: that it acts as a dominant allele in enabling cultured cells to utilize glycerol and as a recessive allele in preventing normal embryo or seed development. This hypothesis was tested by constructing a homozygous diploid from a haploid *Gut* plant obtained from anther culture. All of the progeny produced by the self-fertilization of this plant were viable and gave rise to callus cultures capable of growth on glycerol. At present, therefore, we are searching for another explanation for the aberrant segregation pattern obtained by selfing *Gut/+* heterozygotes. Results of preliminary biochemical experiments indicate that *Gut/+* and normal cells take up ^{14}C-glycerol from the medium at the same rate (R. S. Chaleff and R. L. Keil, unpublished results).

Auxotrophs

Auxotrophic mutants of microbes have been invaluable in defining metabolic sequences and mechanisms of gene regulation. In addition, such mutants are employed routinely in fungal systems for identifying diploids or heterokaryons, and it is anticipated that they will be useful in isolating somatic hybrids formed by fusion of plant protoplasts (see Chapter 5). In microbial and animal cell systems the availability of auxotrophic mutants also facilitates detection of the transfer of isolated DNA (see Chapter 7). Unfortunately, auxotrophs have proved the most difficult class of mutants to isolate in higher plants. As discussed previously, cultured plant cells are not well suited to the selection of nutritional mutants because their requirement for a high minimum inoculum density and their tendency to grow as aggregates (see Chapter 1) make cross feeding unavoidable. Nor is the recovery of recessive mutants favored by the polyploid nature of the genomes of most of those species for which it is possible to regenerate plants from single cells (see this chapter, "Choice of Experimental Material"). Despite these obstacles to the use of selection methods, it should be feasible to isolate auxotrophs simply by screening for nutritional requirements large numbers of clones growing on complete medium.

In the first application of a screening procedure it was discovered that growth of one of several cell cultures derived from *Gingko* pollen was enhanced by arginine (Tulecke, 1960). But since ammonium sulfate and, to a lesser degree, other sources of reduced nitrogen also stimulated growth, it seemed more likely that nitrogen metabolism, rather than arginine biosynthesis specifically, had been altered in this variant cell line.

Recently, a pantothenate-requiring cell line was recovered by screening clones derived from a predominantly haploid cell culture of *D. innoxia*. To produce the large number of clones required for such an endeavor, cell suspension cultures first were filtered until a population composed of more than 60 percent single cells was obtained. Small multicellular aggregates, the majority of which were of single-cell origin, developed during subsequent incubation of this filtered cell suspension. These aggregates then could be plated on supplemented medium at a sufficiently low density to permit the isolation of discrete clones. The one auxotroph was identified among more than 2300 such clones that were tested individually for growth on minimal medium (Savage, King, and Gamborg, 1979).

The task of individually testing thousands or tens of thousands of clones becomes formidable without the assistance of a technique for replica plating, and to anyone who has endured the tedium of this approach, the advantages of selection methods are compelling. However, procedures for selecting auxotrophs are inherently more difficult to design than are those for selecting resistance to an antimetabolite. Generally, in selecting cells that have lost a particular metabolic capability, conditions must be defined that select against normal growing cells while sparing deficient cells that are unable to divide on an unsupplemented medium.

Carlson (1970) first introduced to cultured plant cells a method for selecting auxotrophs that had been developed for use with mammalian cell cultures (Puck and Kao, 1967; Kao and Puck, 1968) and that he had applied earlier to fern gametophytes (Carlson, 1969). This procedure is based on the photolability of DNA into which the thymidine analogue BUdR has been incorporated. When a cell population is incubated in a minimal medium containing BUdR, actively dividing prototrophic cells will incorporate more of the analogue than will auxotrophs. The auxotrophs, therefore, preferentially survive subsequent illumination and can be recovered by transfer to a suitably supplemented medium. Several steps of this mutant selection procedure are critical to its success. First, haploid cells must be used. Carlson obtained cell suspensions from haploid *N. tabacum* plants that had been produced by anther culture. Second, because it is essential that the BUdR be incorporated only by prototrophic cells, it must be added to the medium only after the cessation of DNA synthesis in the auxotrophic cells. Carlson observed that on a minimal medium, cells surviving exposure to a mutagen that killed more than 99 percent of the population and therefore were surely mutant were able to divide twice before depleting their endogenous reserves of metabolites. Considering the generation time of the haploid cells to be approximately 48 hours, the mutagenized population was starved for 96 hours before exposure to BUdR. Presumably, only prototrophs would continue to divide on minimal medium and incorporate the analogue following this initial incubation period. Third, the duration of the exposure to BUdR should be sufficient to kill all of the nonmutant cells. The length of this period was defined by incubating a cell suspension in the dark in the presence of BUdR and, at successive intervals, plating aliquots and then placing these plated cell cultures in the light. As no cells survived illumination after 36 hours of incubation in BUdR-containing medium, suspension cultures were maintained in the presence of the analogue for this length of time before being transferred to a supplemented medium and illuminated. The entire sequence of steps is illustrated in Figure 4.8. From a total population of 1.75×10^6 haploid cells subjected to this procedure, 119 calluses appeared on the supplemented medium. When tested for growth on minimal medium, 6 of these calluses proved to be auxotrophic. The rate of callus growth was restored to normal in three cases by addition to the medium of single amino acids (arginine, lysine, and proline), in two cases by vitamins (biotin and *p*-aminobenzoic acid), and in one by a purine (hypoxanthine). All six auxotrophic cell lines were leaky in that they grew to some degree on unsupplemented medium. Homozygous diploid plants were regenerated from four cell lines and crossed with normal plants to produce heterozygotes. Self-fertilization of plants heterozygous for the mutations causing the biotin, *p*-aminobenzoic acid, and arginine requirements yielded approximately 3 normal progeny to 1 mutant progeny. When heterozygous plants derived from the hypoxanthine-requiring auxotroph were selfed, they produced normal, intermediate, and auxotrophic progeny at approximately a 9:6:1 ratio (P. S. Carlson, personal communication). Therefore, in three cases the auxotrophic phenotype was caused by a single recessive mutation. Inheritance of the hypoxanthine requirement was more complex and may have been caused by two independently segregating recessive mutations that produce an intermediate phenotype in plants homozygous for the mutant allele at either one of the two loci and a fully mutant phenotype in plants homozygous at both loci.

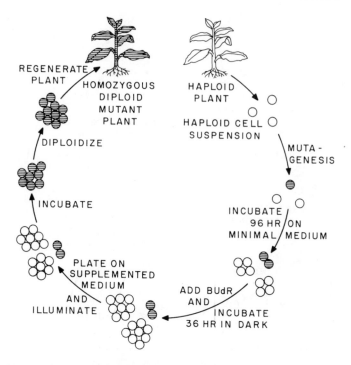

FIGURE 4.8. Schematic representation of the procedure employed by Carlson (1970) to isolate auxotrophic mutants of *N. tabacum*. Cell suspensions derived from a haploid plant were mutagenized and exposed to BUdR. Following transfer to supplemented medium, the cell cultures were illuminated, and the surviving cell lines were isolated and tested for a nutritional requirement.

More recently, a modification of the BUdR selection procedure has been used by Malmberg (1979a) to isolate several temperature-sensitive variant cell lines of *N. tabacum*. Whereas the earlier procedure was based on the inability of the desired conditional lethal variants to grow in the absence of a nutritional supplement and their ability to grow in its presence (Figure 4.8), the modified procedure selected conditional lethal variants on the basis of their ability to grow at a reduced temperature (26°C) but not at an elevated temperature (33°C) (Figure 4.9). Of 84 calluses recovered, 9 that had been obtained from three separate suspension cultures were temperature-sensitive. Two temperature-sensitive variant cell lines of independent origin were characterized more extensively. One cell line *(ts6)* grew normally at 26°C and slowly at 33°C. The second cell line *(ts4)* had a reddish brown color and grew slowly at 26°C, but not at all at 33°C. Growth at the nonpermissive temperature was not restored to either cell line by addition of coconut milk or casamino acids to the

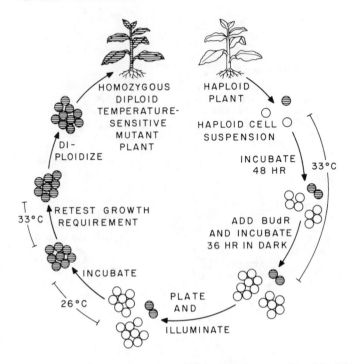

FIGURE 4.9. Schematic representation of the procedure employed by Malmberg (1979a) to isolate temperature-sensitive cell lines of *N. tabacum*. Cell suspensions derived from a haploid plant were incubated at 33°C in a medium containing BUdR. Illumination enriched the population for cells unable to grow at the elevated temperature. These cell lines were recovered at 26°C and then tested for growth at 33°C.

medium. Therefore, either the variants were not auxotrophs or at 33°C they required for growth a metabolite that was not present in the two supplements tested. Plants have been regenerated from both variant cell lines, but at the time of this writing they have not yet flowered. However, *ts4* may yet prove to be a true mutant, since secondary callus derived from plants regenerated from the original cell line manifested the same characteristic reddish brown color and temperature sensitivity (Malmberg, 1979a). Even though it is not yet known whether or not *ts4* is recessive, the isolation of a nonleaky conditional lethal variant of *N. tabacum* contrasts with Carlson's (1970) recovery of only leaky mutants and provides some basis for optimism that nonleaky auxotrophs can be isolated from polyploid species.

Another scheme for selecting auxotrophic variants of cultured plant cells has been proposed by Polacco (1979). This scheme relies on incorporation of the respiratory poison arsenate by metabolically active cells. Twelve colonies (representing a survival frequency of 5.5×10^{-7}) were recovered from a population of mutagenized soybean cells that had been incubated for 1 day in medium containing 2 mM arsenate. Growth of one of these cell lines was reduced on minimal medium and was stimulated by supplementation with proteins or an amino acid mixture. Unfortunately, the variant phenotype reverted to normal before the deficiency could be characterized. But by demonstrating that exposure to arsenate efficiently kills growing cells and does not affect recovery of cells whose metabolism has been arrested reversibly by other inhibitors, Polacco (1979) has illustrated the potential effectiveness of his selection procedure.

Since each selection procedure favors the recovery of a different spectrum of mutants, additional procedures for selecting auxotrophs in plant cell cultures are sorely needed. At present, the best possibilities for use with cultured plant cells appear to be adaptations of procedures for selective elimination of nonmutant cells that have been developed in microbial and animal cell culture systems. One such procedure is based on the observation that on a minimal medium cells with a single growth requirement do not survive a starvation period as well as do cells into which a second metabolic block has been introduced by mutation. In this manner, inositol-requiring strains of *Neurospora* (Lester and Gross, 1959) and yeast (Henry, Donohue, and Culbertson, 1975) and biotin-requiring strains of *Aspergillus* (Macdonald and Pontecorvo, 1953) have been used effectively to isolate auxotrophs by first plating cells in minimal medium, which the singly deficient cells do not survive, and subsequently adding a top layer of complete medium, which rescues cells with multiple deficiencies. A similar system has been developed with mammalian cell cultures by employing aminopterin, a folic acid antagonist that blocks the biosynthesis of thymidylate, to induce a nutritional requirement. Here again, auxotrophs survive starvation better than do prototrophs. This procedure has been used to select glutamine-requiring cell lines from HeLa (DeMars and Hooper, 1960) and Chinese hamster (Chu et al., 1969) cell cultures. A preliminary attempt to apply this procedure to cultured plant cells has yielded encouraging results. It has been shown that sycamore cells that normally require auxin for growth survive exposure to aminopterin much better in an auxin-deficient medium (equivalent to a second nutritional requirement) than in an auxin-supplemented medium (Lescure, 1969). In attempting such experiments, it should be borne in mind that the efficiency of the starvation method for recovering auxotrophs depends on the specific nutritional requirement of the parental strain. Macdonald and Pontecorvo (1953) attributed their success with this procedure to their choice of a biotin-auxotroph of *Aspergillus,* which dies at an exceptionally rapid rate in minimal medium. Since the introduction of an additional nutritional requirement is likely to reduce this death rate, enrichment for the double mutant is achieved.

Another method that may find application to cultured plant cells employs tritiated thymidine to select against prototrophic cells that actively incorporate this radioactive precursor into DNA during growth on a minimal medium. During subsequent incubation at low temperature, prototrophic cells are killed much more rapidly than are auxotrophic cells. By use of this enrichment procedure, populations of *Escherichia coli* cells consisting of more than 90 percent mutant cells have been produced (Lubin, 1959).

Schemes for selecting auxotrophs that rely on selection against growing cells (negative selection schemes) generally are difficult to implement because they require precise definition of several experimental parameters. Wherever possible, it is far easier to select variants on the basis of their ability to grow under conditions that inhibit growth of normal cells, but thus far, few such positive selection schemes have been devised for isolating auxotrophs. In one case it was shown that bacterial mutants requiring thymine can be selected on medium supplemented with thymine and trimethoprim, an inhibitor of dihydrofolate reductase (Stacey and Simson, 1965). However, some care must be exercised in applying microbial selection schemes to cultured plant cells. Resistance to methyl mercury and growth on medium containing α-aminoadipic acid as a nitrogen source have been employed to select auxotrophic mutants of yeast requiring methionine (Singh and Sherman, 1974) and lysine (Chattoo et al., 1979), respectively. But because yeast and higher plants accomplish the synthesis of these amino acids by quite different pathways, it is doubtful that these selection methods will be useful with cultured plant cells. Of course, auxotrophs obtained by positive selection methods can be used in reconstruction experiments to define optimal conditions for recovering additional auxotrophs by other means. The availability of several selection schemes will greatly increase the number of different classes of mutations of a particular phenotype that can be isolated.

Increased tolerance for herbicides and fungicides

The effectiveness of herbicides is based on their ability to discriminate between the weed and crop species. Traditionally this differential has been achieved by identifying compounds that display such specificity and therefore has been regarded as being solely the province of chemists. Alternatively, however, differential responses to a particular herbicide can be introduced by enhancing resistance of the crop species by genetic means. This genetic approach should broaden the spectrum of applicability of existing herbicides and thereby spare the enormous expense of developing and licensing new herbicides.

Cell lines tolerant for several phenoxy herbicides [2,4-dichlorophenoxyacetic acid (2,4-D), 2,4,5-trichlorophenoxyacetic acid, and 4-(2,4-dichlorophenoxy) butyric acid] have been selected from suspension cultures of *Trifolium repens* (Oswald, Smith, and Phillips, 1977). In addition, a *N. sylvestris* cell line resistant to 2,4-D has been reported (Zenk, 1974). Several cell lines have also been recovered from the plating of *N. tabacum* protoplasts in the presence of a toxic concentration of the herbicide isopropyl-*N*-phenylcarbamate. But none of the regenerated plants has displayed a clearly increased tolerance for the herbicide, nor have any produced resistant seed (Aviv and Galun, 1977). Shoots have been regenerated from aminotriazole-resistant *N. tabacum* cell lines as well (Barg and Umiel, 1977). However, to my knowledge, no description of the characteristics of these shoots or of genetic studies has been published.

The feasibility of selecting from among cultured cells genetically stable mutations that enhance tolerance of the whole plant for a particular herbicide has been demonstrated by the isolation of picloram-resistant mutants of *N. tabacum*. Of seven cell lines initially selected by plating diploid cell suspension cultures on 500 μM picloram, plants regenerated from five gave rise to resistant secondary callus cultures. On a medium containing 100 μM picloram, limited seedling development occurred from seeds produced by plants regenerated from four herbicide-resistant cell lines, whereas no such development occurred from normal seeds plated on this medium. These differential responses of normal and mutant seeds to picloram greatly facilitated the scoring of crosses with plants regenerated from these four cell lines and provided the first evidence of expression of genetic herbicide tolerance in the plant. Backcrosses and self-fertilization of plants regenerated from three picloram-resistant cell lines produced resistant and sensitive progeny at the ratios of 1:1 and 3:1, respectively. Seedlings developed from approximately one-half of the seeds obtained from self-fertilization of plants regenerated from the fourth selected cell line (the R_1 generation; see this chapter, "A matter of terminology") displayed an intermediate degree of resistance to picloram, whereas one-quarter were strongly resistant, and another one-quarter were sensitive. These results were amplified by additional analysis of the R_2 generation (progeny resulting from selfing of R_1 plants) of one mutant. Thus, of the four cases in which picloram resistance was transmitted across sexual generations, three mutations (*PmR1, PmR2,* and *Pm-R7*) behaved as dominant alleles, and one (*PmR6*) behaved as a semi-dominant allele of single nuclear genes. In one case (*PmR1*) the complete dominance of the resistance phenotype was confirmed by growth tests

Table 4.7. *Determination of linkage among* PmR *alleles*

Phenotypes of plants crossed	No. of resistant individuals		No. of sensitive individuals	
	Observed	Expected	Observed	Expected
PmR1; *PmR7* × +	426		0	
PmR1; *PmR7* selfed	244		0	
+ × *PmR7*; *PmR6*	162	174	70	58
PmR7; *PmR6* selfed	228	236	24	16
+ × *PmR1*; *PmR6*	201	197	62	66
PmR1; *PmR6* selfed	240	241	17	16
PmR7; *PmR85* × +	213	224	86	75
+ × *PmR7*; *PmR85*	261	253.5	77	84.5
PmR7; *PmR85* selfed	538	533	31	36
PmR6; *PmR85* × +	256	246	72	82
+ × *PmR6*; *PmR85*	254	250.5	80	83.5
PmR6; *PmR85* selfed	230	238	24	16

Source: R. S. Chaleff (1980a).

of callus cultures initiated from plants of the three different genotypes segregating in the R_1 (+/+, *PmR1*/+, *PmR1*/*PmR1*). Because growth of callus cultures established from homozygous mutant plants (*PmR1*/ *PmR1*) was inhibited to nearly the same extent (66 percent) by 1000 μM picloram as was that of normal callus by only 10 μM (69 percent), it is apparent that this mutation increases cellular tolerance for the herbicide 100-fold. Since these experiments were performed with diploid cells, the recovery of only dominant and semidominant mutations was to be expected (Chaleff and Parsons, 1978b). Plants regenerated from another independently isolated picloram-resistant cell line have also been shown to be heterozygous for a semidominant nuclear mutation (*PmR85*) (R. S. Chaleff, unpublished results). Uptake of ^{14}C-picloram was unimpaired in seedlings homozygous for *PmR1*, *PmR6*, *PmR7*, and *PmR85*, and no difference was detected between normal and mutant plants in metabolism of the herbicide (Chaleff, 1980a). These homozygotes were crossed to construct heterozygotes, which then could be crossed with a normal sensitive plant and selfed to test for linkage of the *PmR* mutations. In the case in which two mutations were linked, only resistant progeny could be obtained from both self-fertilization and a testcross. However, if two mutations are unlinked, self-fertilization of the double heterozygote should produce a 15:1 ratio between resistant and sensitive progeny, and one-fourth of the progeny of a testcross should be sensitive. It is evident from the data presented in Table 4.7 that *PmR1* and *PmR7* are linked, whereas *PmR6* and *PmR85* are located in distinct linkage groups. Since

FIGURE 4.10. Effects of picloram on normal (top) and homozygous mutant (*PmR1/PmR1*) (bottom) tobacco seedlings. Seedlings were grown axenically for 8 weeks. Picloram was then added to several of the beakers to the following final concentrations: no picloram (left), 1 μM (center), and 5 μM (right). (R. S. Chaleff, unpublished results.)

PmR1 and *PmR7* were derived from the same experiment, these two isolates may not have arisen independently. A comparison of the responses of normal and *PmR1/PmR1* seedlings to exposure to picloram clearly demonstrates the expression of tolerance in mutant plants (Figure 4.10).

Although selection at the cell level has been used successfully to isolate mutant plants exhibiting enhanced tolerance for picloram, the applicability of this procedure is obviously limited. One would expect such an approach to work with herbicides that interfere with basic metabolic functions expressed by cultured cells, but not with those that disrupt more specialized functions found only in differentiated cells of a mature plant. For example, herbicides that inhibit photosynthesis might not even prove toxic to cultured cells. Such a constraint is treated more thoroughly in Chapter 2. But a method has been introduced by Radin and Carlson (1978) that enables us to step beyond these confines of the Petri dish. Although

Table 4.8. *Segregation of Bentazone resistance and Phenmedipharm resistance among progeny of heterozygous F_1 plants.*

Mutant isolate from which F_1 plant was derived	No. of F_2 individuals		Proposed ratio
	Resistant	Sensitive	
Bentazone-resistant			
B-1	9	91	1:15
B-2	21	79	1:3
B-3	33	67	1:3
B-4	12	88	1:15
B-5	29	71	1:3
B-6	4	96	1:15
B-7	19	81	1:3
B-8	20	80	1:3
Phenmedipharm-resistant			
P-1	23	77	1:3
P-2	16	84	1:3

Source: Radin and Carlson (1978).

the two herbicides Bentazone and Phenmedipharm do not affect growth of *N. tabacum* callus cultures, they do bleach the leaves of intact plants. If resistance were to be found, it had to be sought in situ among the populations of differentiated leaf cells that are sensitive to these herbicides. Haploid tobacco plants produced by anther culture were mutagenized by exposure to γ-irradiation. After these plants were sprayed with the herbicides, occasional green islands of resistant cells were visible in the otherwise yellow leaf tissue. These sectors were excised, placed into culture, and induced to form plants. Regenerated plants were diploidized and crossed to a normal plant. Although all of these heterozygous F_1 plants were sensitive to the herbicides, self-fertilization produced herbicide-resistant progeny among the F_2 generations of eight Bentazone-resistant and two Phenmedipharm-resistant isolates. The segregation ratios indicated that both cases of Phenmedipharm resistance and five cases of Bentazone resistance were due to single recessive mutations. Resistance of three other isolates to Bentazone apparently was caused by two independently segregating recessive mutations (Table 4.8). These mutations constitute several different complementation groups. Although the in situ mutant selection procedure employs cell culture only to induce the formation of plants from excised mutant leaf tissue (and therefore any mention of it in this discussion might seem inappropriate to some), it is presented here as an illustration both of the limitations of cell culture in isolating plant mutants and of an innovative means of transcending those limitations.

Also of potential value is the improvement of plant tolerance for pesticides and fungicides. After demonstrating that the fungicide carboxin inhibits both growth of tobacco callus and oxidation of succinate by tobacco mitochondria, Polacco and Polacco (1977) had reason to believe that mutants selected for resistance to carboxin would possess an altered form of succinate dehydrogenase. But "the best laid schemes o' mice an' men gang aft agley." A cell line resistant to carboxin was isolated from cell suspension cultures derived from a haploid *N. tabacum* plant. However, succinate oxidation by mitochondria extracted from normal and resistant cells is equally sensitive to inhibition by carboxin, as is respiration of intact normal and resistant cells growing on sucrose. Therefore the biochemical basis of resistance of this cell line to carboxin is not yet known. The regeneration of fertile plants permitted genetic studies to be performed. All progeny obtained from self-fertilization of these regenerated plants were resistant to carboxin. These resistant progeny were crossed reciprocally with normal plants, and again only resistant progeny were recovered. However, among the F_2 progeny were 25 resistant and 15 sensitive seedlings. Unfortunately, as in other cases that have been considered, the number of progeny that could be scored was limited because the mutant phenotype could be detected only in callus cultures and not in plants. Hence, scoring of crosses must proceed by determining the response to carboxin of callus cultures initiated from progeny seedlings. On the basis of the available data it can be said that resistance is dominant, but whether it is caused by one or two nuclear mutations remains to be determined by further analysis (J. C. Polacco, personal communication).

Tolerance for environmental stress

Improving the tolerance of plants for environmental stresses is of such paramount importance that it is inevitable that the applicability of cell culture to this purpose be explored. Selection of plant cells resistant to drought and chilling injury, salts, and heavy metals is made difficult by the present inadequacy of our knowledge about the responses of plants to these stresses and the means that plants might develop for coping with them. But perhaps the selection of appropriate variants will provide tools for acquiring the basic knowledge now lacking.

The most obvious method of selecting cell lines with enhanced chilling tolerance is to place cultures at a temperature that kills the majority of cells and isolate those callus pieces that grow when the cultures are returned to 25 °C. In one such experiment, Petri dishes that had been inoculated with cell suspension cultures established from haploid and diploid *N. sylvestris* and diploid *Capsicum annuum* plants were incubated at 5°C (for *C. annuum*) or 0°C (for *N. sylvestris*) for a period of 21 days. Following transfer to 25°C, a few colonies formed on many plates. But in most cases the numbers of colonies appearing on plates that had been refrigerated were more than 1 percent of the numbers forming on plates that had been maintained at 25°C. This frequency is much higher than that expected for a mutational event. Nevertheless, several of these selected cell lines, when retested, continued to display greater capacity than the parental cell lines to recover from exposure to low temperature (Dix and Street, 1976). Plants were regenerated from three such cell lines of *N. sylvestris,* but callus cultures initiated from progeny of these regenerated plants were as sensitive as normal cells to cold treatment (Dix, 1977). Apparently the chilling tolerance of the selected cell lines was due either to adaptation or to an epigenetic change (see Chapter 2).

In another attempt to select cold-tolerant cells, spruce callus cultures first were incubated at 2°C and then were transferred progressively to lower temperatures. Very few callus pieces survived a 24-hour exposure to −20°C. However, half of the callus tissue of one variant cell line recovered from exposure to −35°C (Tumanov et al., 1977). Unfortunately, it was not possible to regenerate plants from this cell line for further studies.

Cell lines capable of growth in the presence of a normally toxic concentration of NaCl have been isolated from cell cultures of *N. sylvestris* (Zenk, 1974; Dix and Street, 1975), *C. annuum* (Dix and Street, 1975), *N. tabacum* (Nabors et al., 1975), and alfalfa (Croughan, Stavarek, and Rains, 1978). A study of the intracellular concentrations of various ions revealed that during growth in the presence of NaCl, variant alfalfa cells retain potassium ions more effectively than do normal cells. At low salt concentrations variant cells accumulate more nitrate than do normal cells, although at higher salt concentrations the nitrate contents of both types of cells are similar. Optimal growth of this variant alfalfa cell line occurs on a medium supplemented with 0.5 percent (86 mM) NaCl. Plants have been regenerated from a salt-tolerant diploid cell line of *N. tabacum*. These plants were able to withstand irrigation with a 560 mM NaCl solution, which is higher than the NaCl concentration of seawater. Self-fertilization of these plants produced a larger proportion of salt-tolerant progeny than was found among progeny of plants regenerated from normal cells. Nearly all of the progeny of the next generation were salt-tolerant. However, if this tolerance were due to a single dominant nuclear allele, more, not fewer, sensitive plants would be expected in the R_2. Additional crosses, including backcrosses, must be performed in order to establish the basis of the tolerance of these plants for NaCl. The reported results do not eliminate the possibility that tolerance is due to adaptive changes (such as an altered membrane composition or enzyme levels) that are transmitted maternally in crosses (Nabors et al., 1980). If the salt tolerance of these tobacco plants proves to be genetic, the practical value of plant cell culture could be contested no longer.

Selection of cell lines resistant to heavy metal ions would seem to be a straightforward matter, but experiments are complicated by the pH dependence of the solubility of many metal ions. For this reason, studies of the effects of aluminum on plants are performed below pH 5, where aluminum is soluble. But cultured plant cells cannot grow in this pH range. At higher pH values aluminum ions are essentially removed from the medium by formation of $Al(OH)_3$, which is insoluble. Meredith (1978) surmounted this dilemma by incorporating a chelating agent (EDTA) in the culture medium. By this method aluminum ions were made soluble in the mildly acidic pH range favorable to plant cell growth in vitro, and tomato cell lines resistant to a toxic concentration (200 μM) of chelated aluminum could be selected. In such experiments there is the danger that the chelating agent itself may prove toxic by sequestering other cations essential for cell growth. Therefore, appropriate controls must be performed to demonstrate that toxicity is actually due to the metal ion.

Determination of frequency and rate of mutation

The rates and frequencies of forward and reverse mutations are fundamentally important properties of any genetic system. Quantitative studies of these parameters are needed to optimize the recovery of additional mutations and to compare the efficiencies of different mutagens. The frequencies of reversions in response to different mutagens also assist in the characterization of mutations. For example, the failure of a particular mutation to revert provides the first indication that it is a deletion. And frameshift mutations can be identified tentatively by their reversion in response to acridines, rather than to mutagens that induce base substitutions (as reviewed by Hayes, 1970). But possibly the greatest significance of reversion rate and frequency to plant cell genetics is their use in ascertaining the stability of mutations and hence their value as genetic markers. Without knowledge of the spontaneous reversion frequency of a marker mutation, it cannot be presumed that phenotypic changes observed in a particular experimental system result from the transfer of genetic material by protoplast fusion or DNA or organelle uptake rather than simply from the reversion of that mutation.

In experiments with microbes, the number of mutations of a particular type present in a population is determined most directly by plating a large population of cells on selective medium and counting the number of mutant colonies that appear. Division of this number by the total number of viable cells in the plated population gives the mutation frequency. But the performance of such an experiment with cultured plant cells is not as simple, because of several features of the growth of these cells in vitro.

First, let us consider the difficulties of determining the number of viable cells in a population of plant cells. To determine the density of a microbial population, a series of dilutions of the cell suspension is prepared, and aliquots of these dilutions are plated. The numbers of colonies are counted on plates on which they develop as discrete units. From the size of the aliquot plated and the dilution made, it is then a simple matter to calculate the number of viable cells in the original population. But the requirement of a minimum inoculum density for growth of plant cells in culture (see Chapter 1) rules out this method. One could consider plating aliquots of dilutions on a conditioned medium, but the proportion of cells growing under these conditions might be very different from the proportion that would grow when plated at a high density on a minimal medium. The microbial methods also assume the availability of single cells. Therefore, although these methods might be feasible with plant protoplast preparations, they would not be so with cultured cells, which tend to grow as aggregates. Such aggregates also prevent determination of population size by direct counting of the number of cells in a suspension aliquot. This problem has been dealt with in a number of ways. One method employs chromic acid to dissociate cell aggregates (Yeoman, 1973). Other methods estimate the size of the population from its weight or volume. But such procedures permit determination of only the total number of cells in the population rather than the number of cells capable of forming colonies. Efforts to measure the number of growing cells in a suspension culture on the basis of its turbidity (Sung, 1976) are unreliable because of the influence of aggregate size on light scattering and the exudation by growing cells of macromolecules and metabolites that contribute to the opacity of the medium. Another complication in determining population size is that it can change during the course of the experiment. If the selection medium retards, rather than prevents, growth of nonmutant cells, the population size is increasing constantly. Since mutant cells will arise from division of nonmutant cells, under such conditions one is measuring not the mutation frequency (the number of mutations present in the population) but the sum of the mutation frequency and the mutation rate (the number of mutations occurring per cell per generation).

Now let us consider some problems associated with determining the number of mutations occurring in the population. In most experiments only a fraction of the cell lines isolated after selection have continued to display an altered phenotype on retesting, and therefore many of the initial isolates are most likely not true mutants. Accordingly, simply counting the number of colonies appearing on the selection medium is a meaningless gesture. One factor that contributes to this inflation of the number of surviving cells is the release by dying cells of metabolites that assist other cells in overcoming the conditions of selection. Yet other cells may be protected by their location in the midst of an aggregate through which the selection agent cannot penetrate. Of course, when retested under more stringent conditions, such cell lines will display a normal phenotype. Another cause of phenotypic instability is the occurrence of epigenetic events (see Chapter 2). Epigenetic changes may revert at a high rate. However, such changes may also be relatively stable, in which case, even after retesting the initial isolates, one may be measuring not the frequency of mutation but the frequency of cells displaying an altered phenotype as a result of both genetic and nongenetic events. The relative frequencies of genetic and epigenetic changes may vary with each system. Therefore, before employing the appearance of a phenotype as a measure of mutation frequency or rate, it is imperative that a sufficient number of isolates be characterized by genetic analysis of regenerated plants to discover the proportion of variant isolates that are truly mutant.

As an alternative to measuring the mutation frequency, the mutation rate can be determined. Two methods for estimating mutation rates that are perhaps most familiar and most readily adaptable to plant cell cultures are the statistical treatments devised by Luria and Delbrück (1943) of the results of fluctuation tests. In such experiments many independent flasks containing liquid medium are inoculated with equal numbers of cells. It is critical that the inocula be sufficiently small that mutations do not arise in a fraction of the cultures. In the first method, the average number of mutations per culture is calculated according to Poisson's formula from the proportion of cultures containing no mutants. The mutation rate is then computed from this number and the final number of cells in each flask. In the second method, the mutation rate is calculated from the average number of resistant cells and the final population size per culture. Both of these methods for calculating the mutation rate also require that the number of cells in the culture be counted. But as it is the final size of the population that must be determined, one does not need to be as concerned about the proportion of the cell population that is viable, since this population is the product of active cell division. Therefore, the number of cells in the final population can be determined by the methods discussed previously, such as dissociating aggregated cells in an aliquot of the culture or filtering the culture and weighing the washed cells. Of course, in measuring mutation rate one must exercise the same cautions in making assumptions about the basis of an altered phenotype that were discussed in reference to the measurement of mutation frequency.

A matter of terminology

As no consistent terminology has yet been applied to the description of crosses involving plants regenerated from cultured cells, such description often becomes awkward and prolix. Diploid mutant plants regenerated from cultured cells can be either heterozygous, if the mutation initially occurred in a diploid cultured cell, or homozygous, if the mutation occurred in a haploid cell that subsequently became diploid through endomitosis or endoreduplication or if a heterozygous diploid cell became homozygous by mitotic recombination. Accordingly, the genotype of a regenerated plant depends on its history in vitro, and this genotype will determine in which generation following crosses with such a plant genetic segregation will first appear. Quite understandably, those who are accustomed to seeing segregation first in the F_2 and never in the F_1 or P_1 (*vide infra*) generations are distraught by any reference to the segregating progeny of a heterozygous regenerated plant as an F_1.

Problems of nomenclature inevitably arise from the introduction of new techniques or concepts, as they did when it first became necessary to devise terms for the progeny of a controlled series of crosses. At that time Bateson and Saunders (1902) wrote:

It is absolutely necessary that in work of this description some uniform notation of generations should be adopted. Great confusion is created by the use of merely descriptive terms, such as "first generation," "second generation of hybrids," etc., and it is clear that even to the understanding of the comparatively simple cases with which Mendel dealt, the want of some such system has led to difficulty. In the present paper we have followed the usual modes of expression, but in future we propose to use a system of notation modelled on that used by Galton in 'Hereditary Genius'. We suggest as a convenient designation for the parental generation the letter P. In crossing, the P generation are the pure forms. The offspring of the first cross are the first filial generation F. Subsequent filial generations may be denoted by F_2, F_3, etc. Similarly, starting from any subject-individual, P_2 is the grandparental, P_3 the great-grandparental generation, and so on.

Clearly, most of this now conventional terminology applies to plants regenerated from cultured cells. But the difference is that according to this notation, parents of an F_1 generation are always pure (i.e., homozygous), whereas plants regenerated from cultured cells are not. What notation is appropriate to denote the progeny of selfs or crosses with regenerated plants of unknown genotype? Even this problem of a terminology for crosses with heterozygotes was discussed by Bateson and associates in 1905: "'F_1' throughout designates the first filial generation resulting from a cross. 'F_2' means the generation resulting from F_1, and, unless qualified, means the offspring of $F_1 \times F_1$. Other such terms are urgently needed, e.g., for the various offspring of DR $\times$ R, etc."

Accordingly, I would like to propose that plants regenerated from cultured cells, whether they be homozygous or heterozygous, be designated R (for regenerated plant). Self-fertilization of regenerated plants will produce the R_1 generation; self-fertilization of R_1 plants will produce the

R_2 generation, and so forth. Thus, if a regenerated plant (R) is heterozygous, segregation will appear in the R_1 generation, and in the case that R is homozygous, segregation will not occur in either the R_1 or subsequent R generations. Crosses of regenerated plants with homozygous plants (P) can be designated R × P. Progeny of such a cross should be referred to as the F_1, and, employing the established convention, progeny resulting from self-fertilization of the F_1 should be referred to as the F_2, and so forth. If R is heterozygous, segregation will occur in the F_1, and if R is homozygous, segregation will not occur until the F_2.

5
Protoplast fusion

For several years cell fusion remained a technique to be applied exclusively to animal cells and to be eyed enviously by those grappling with plant cells, for until the development by Cocking (1960) of a technique for its enzymatic digestion, the cell wall had protected the intact plant cell from the encroachments of even the most eager experimentalists. However, with the exposure of the plasmalemma, the impregnability of the cell was lost, and a variety of new genetic manipulations became possible. Among the operations that then became feasible were protoplast fusion and the uptake of organelles (see Chapter 6) and of DNA (see Chapter 7). An additional advantage of the enzymatic procedure was that it made available large populations of genetically and physiologically homogeneous protoplasts. Moreover, these protoplasts were free of intercellular bonds and involvements and could be regarded as genetically and developmentally independent units.

Perhaps the most universally recognized experimental opportunity introduced by the recently developed ability to fuse protoplasts is the hybridization of widely divergent taxa that normally are isolated genetically by morphological, developmental, or sexual incompatibility barriers. By this method of cell fusion, new and otherwise impossible genetic combinations can be constructed. Somatic hybridization might also be exploited as a rapid means of forming heterozygous diploid cells to accelerate the genetic characterization of plants. It may become feasible to perform genetic mapping by mitotic recombination as in certain fungal systems and analysis of gene linkage by techniques that have become routine with animal cell cultures (Pontecorvo, 1958).

In addition to the union of nuclear genomes, protoplast fusion also accomplishes the coalescence of cytoplasms. The mixing of cytoplasms is not usually possible, because in only a few plant species is the cytoplasm transmitted through both the male and female gametes. The ability to juxtapose nuclei, chloroplasts, and mitochondria in new combinations within a common cytoplasm provides a novel and potent method to investigate the relationships of these organelles. The possibility of performing complementation and recombinational analyses of functions encoded by the mitochondrial and chloroplast genomes is suddenly

96

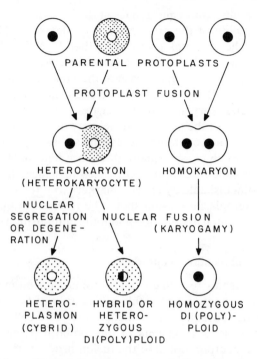

FIGURE 5.1. Schematic representation of events that can occur during protoplast fusion and their products. Although the fusion of genetically identical
protoplasts is illustrated for only one parental type, the same process can occur
for protoplasts of the other parental type. Diploids are produced by fusion of two
haploid protoplasts as shown. However, polyploids can be produced by fusion of
nonhaploid protoplasts or of more than two protoplasts.

brought nearer to reality. The vegetative transfer of extrachromosomal
characteristics, such as cytoplasmic male sterility, a valuable trait for
hybrid seed production, also becomes conceivable. And our understanding of the control of differentiation and gene expression should be
expected to benefit from the study of hybrids formed between cells in
distinct states of differentiation.

Products of protoplast fusion

The events involved in somatic cell fusion are depicted in Figure 5.1. First
the plasmalemmas fuse to form a single cell in which the parental nuclei
remain separated within a mixed cytoplasm. This first product, in which
cells of different genotypes have fused but the nuclei have not, is called a
heterokaryon (the term heterokaryocyte is sometimes employed). If
genetically identical protoplasts fuse and karyogamy does not occur, a

homokaryon results. Subsequent nuclear fusion will produce a synkaryon. The final ploidy of the products will be determined by the number and ploidy of the parental protoplasts that fuse. In the case of the heterokaryon, the product of karyogamy is a hybrid or heterozygous diploid (or polyploid), and in the case of the homokaryon, a homozygous diploid (or polyploid) is formed. If nuclear fusion does not occur, the nuclei of a heterokaryon may segregate into separate cells during cytokinesis, or one of the parental nuclei may degenerate. Either process will create a cell that contains the cytoplasm of both parents, but the nucleus of one. Such cytoplasmic hybrids are heteroplasmons, but they have also been called "cybrids." In cases in which the parental cytoplasms are distinguishable by genetic markers, but the nuclei are not, the products of heteroplasmic fusion should be referred to by these terms to avoid any unwarranted assumption about the composition of the nucleus.

Karyotypic instability of fusion products

The genetic amalgams resulting from protoplast fusion may be produced by the equal combination and stable association of two parental nuclei. However, during subsequent divisions of the hybrid cells, whole chromosomes or chromosome fragments of one or both parents may be lost, resulting in the incorporation into the fusion product of a fraction of one or both parental genomes. Such events can generate a variety of aneuploids and chromosomal insertions or deficiencies. Two factors determining the genetic stability of the hybrid are the two species being combined and the length of the period that it is maintained in culture.

Any study of karyotypic instability of somatic hybrids must discriminate between changes in chromosome numbers and composition caused by the inherent genetic instability of the cell culture system itself (see Chapter 3) and changes arising from the unacceptability of a strange partnership forced on two incompatible genomes in the hybrid cell. The source of any observed karyotypic variation may be identified through knowledge of the chromosome composition of the two parental protoplasts and by monitoring changes in these karyotypes during propagation of the hybrid. The two possible origins of genetic instability were not distinguished in the characterization of somatic hybrids formed by the fusion of protoplasts produced from leaf mesophyll cells of *Nicotiana glauca* and protoplasts obtained from cell suspension cultures of *Nicotiana tabacum*. Chromosome numbers of plants regenerated from the somatic hybrid cultures tended to be higher than the sum of the two parental karyotypes, and enormous variation in these numbers was observed among cells within the same root tip (Maliga et al., 1978). But these higher chromosome numbers could have been generated in the *N. tabacum* suspension cultures and detected only when the chromosomes of the fusion product were counted. Nonspecific changes in chromosome composition occurring during maintenance of the hybrid cells in culture might have generated a heterogeneous cell population from which chimeral root tips could have originated.

Although soybean protoplasts used in the construction of soybean + *N. glauca* hybrids were obtained from cell suspension cultures, the structural instability and rapid loss of the *N. glauca* chromosomes left little doubt that the observed karyotypic changes were a peculiar property of the hybrid, a phenomenon well known to students of animal cell culture. After 6 months a few tobacco chromosomes were still visible in the somatic hybrids, and these remaining tobacco chromosomes then appeared to be dividing synchronously with the soybean chromosomes. The occurrence of chromosomal bridges at mitosis suggested that rearrangements had occurred as well (Kao, 1977). Loss of fragments of the *N. glauca* genome was also suggested by the disappearance from the hybrid cell lines of tobacco alcohol dehydrogenase and aspartate aminotransferase isozymes over a period of 8 months (Wetter, 1977).

An intriguing example of preferential loss of the chromosomes of one parent in somatic hybrids is offered by studies of cell lines produced by the fusion of *Petunia hybrida* and *Parthenocissus tricuspidata*. Although the hybrid cells did not appear to contain any of the morphologically distinct *Petunia* chromosomes, they did possess isoperoxidase isozymes of both parents. The results suggested that the *Petunia* chromosomes were largely excluded from the hybrid, but that the genes coding for the *Petunia* isoperoxidases were retained and may have been integrated into the *Parthenocissus* genome (Power et al., 1975). Likewise, chromosomes of plants regenerated following the fusion of protoplasts of *Daucus carota* and *Aegopodium podagraria* were similar in number and morphology to those of carrot ($2n = 18$), but since these plants synthesized chlorophyll and the carrot parent was albino because of nuclear mutation, it appeared that not all of the *Aegopodium* genes had been eliminated. Another *Aegopodium* trait expressed by the regenerated plants was the production of neurosporene in the roots. The results of molecular hybridization experiments also suggested that *Aegopodium* genes had been incorporated into the nuclear genome of the regenerated plants. Although carrot nuclear RNA hybridized to a small degree with *Aegopodium* DNA, a significantly greater amount of hybridization occurred between *Aegopodium* DNA and nuclear RNA extracted from the regenerated plants. Thus, although the genome of the regenerated plants was cytologically indistinguishable from that of carrot, evidently *Aegopodium* genes were being transcribed in the nuclei of these plants (Dudits et al., 1979).

But specific chromosome elimination is not necessarily the rule. Nor can its occurrence be predicted by the evolutionary distance between the two parents. In somatic hybrids produced by the fusion of *Arabidopsis thaliana* and *Brassica campestris,* some rearrangements were evident after 5 months of culture, but the complete chromosome sets of both parents were retained, and changes in chromosome numbers were not observed until after 7 months (Gleba and Hoffmann, 1978).

As the drastic karyotypic changes that can occur during the initial culture period of some somatic hybrids are characterized by the preferential loss of one parental genome, they differ from the apparently more random changes that are observed in other cell cultures (see Chapter 3). An explanation for this preferential chromosome loss might be found in a consideration of the trials faced by a hybrid genome. Perhaps the most severe problem confronting the somatic hybrid is how to establish harmony between two genomes that have evolved separately and most probably encode conflicting programs for the processes of cell division, development, and differentiation. Even if the two genomes can be synchronized in such basic activities as DNA replication and karyokinesis, the coordination of developmental sequences and the integration of specialized and possibly antagonistic structures and functions must be accomplished. For example, how would the meristematic centers be organized in a hybrid between a monocot and a dicot? Conciliation between divergent cellular and developmental programs of two parents may be realized easily by the simple dominance of one parental set of regulatory signals. But it is easier to imagine a rapprochement being realized through the introduction into one parental genome of only functionally compatible fragments of the other. During subsequent cell divisions, selection probably would favor cells containing whole chromosomes or pieces of chromosomes of one parent that can coexist with the genome of the other, while cells possessing more discordant genetic information would be excluded.

Numerous fusions have been accomplished between protoplasts of evolutionarily distant origins. Many of these hybrids did not proceed beyond a few cell divisions, whereas others developed into stable cell lines apparently capable of dividing indefinitely. But those hybrid cell lines from which plants could not be regenerated lie for the moment beyond genetic analysis and therefore also beyond the scope of this volume. For information about these achievements, the reader is referred to recent reviews by Constabel (1976, 1978) and Gamborg (1976, 1977).

Selection of somatic hybrids

In recent years several techniques have been developed for inducing efficient fusion of plant protoplasts. These methods include the use of polyethylene glycol (Wallin, Glimelius, and Eriksson, 1974; Kao and

Michayluk, 1974) and incubation in an alkaline medium containing a high concentration of calcium ions (Keller and Melchers, 1973). The principal difficulty involved in the successful isolation of somatic cell hybrids is the development of means for identifying the hybrid and distinguishing it from the parental cell types.

Although fusion occurs readily, hybrid cells have been recovered at but a low frequency in the successful fusion experiments reported to date. Visual and cytological markers have served to identify a fusion event, but because protoplasts must be cultured at a high density (approximately 10^4 protoplasts per milliliter) to enable them to divide and form a callus, unless the hybrids possess an advantage over the parental cells they will be overgrown rapidly by a proliferating mass of parental callus. Efficient recovery of fusion products requires the definition of selective culture conditions that permit growth of the hybrid and inhibit or preclude growth of the parental cells. The design of a selective system is greatly facilitated if a sexual hybrid between the two species can be generated. Because the sexual hybrid should display the same properties as the expected somatic hybrid, a reconstruction experiment can be performed in which the growth requirements and natural resistances to antimetabolites of proto- plasts from the sexual hybrid and the two parental species can be compared. Such a procedure may reveal conditions that favor growth of hybrid protoplasts and suppress growth of parental protoplasts. Of course, sexual hybrids will not always be available. Indeed, as the techniques for somatic hybridization improve, it is hoped that protoplast fusion will be applied for the purpose of constructing hybrids that cannot be produced by sexual crosses. But in those cases in which sexual hybrids are available, they are a valuable resource for defining optimal conditions for recovery of the somatic hybrid. Let us now consider the three different types of selective systems that have been employed to this end.

Selections based on natural characteristics

The earliest application of a selection procedure to the recovery of protoplast fusion products yielded hybrids of *N. glauca* and *N. langsdor- fii*. This procedure was based on the inability of protoplasts of *N. glauca* and *N. langsdorfii* to form callus on a medium that supported division of a small proportion of the hybrid protoplasts and on the ability of hybrid callus to grow in the absence of exogenous hormones that were required for growth by callus of the two parental species (Carlson, Smith, and Dearing, 1972). The task of defining these selective growth conditions was made easier by the availability of the amphidiploid hybrid from sexual crosses. From a mixed population containing more than 10^7 protoplasts of each parent, 33 calluses were recovered, and all grew vigorously following transfer to a medium lacking hormones. Characterization of three isolates from which plants were regenerated verified their hybrid genotype. The regenerated shoots produced leaves that were morphologically indistinguishable from those of the sexual hybrid and distinct from those of either parent. The density of trichomes on the leaves of both the somatic and sexual hybrids was intermediate between that on the pubescent *N. langsdorfii* and that on the glabrous *N. glauca*. The somatic hybrid also possessed the same capacity for spontaneous tumor formation that is peculiar to the sexual hybrid. The chromosome number of the somatic hybrid (42) was the sum of the diploid somatic numbers of *N. glauca* (24) and *N. langsdorfii* (18). The leaf peroxidase isozyme banding pattern of the somatic hybrid was identical with that of the sexual hybrid and represented the combined isozyme bands of the two parental species. In fact, the somatic hybrid differed from the sexual hybrid only in being self-fertile. Additional confirmation of the hybrid genotype of the regenerated plants was provided by analysis of the subunit polypeptide composition of ribulose bisphosphate carboxylase of plants grown from seed produced by progeny of the somatic hybrid (the F_3, if the original somatic hybrid is counted as the F_1). By means of isoelectric focusing in 8 M urea, the nuclear-encoded small subunit and the chloroplast-encoded large subunit of this enzyme can be resolved into their component polypeptides (Kung, Sakano, and Wildman, 1974). The patterns obtained are characteristic of each species and therefore serve as markers for the nuclear and chloroplast genomes. These experiments revealed that the descendant of the somatic hybrid contained the small-subunit polypeptides of both *N. glauca* and *N. langsdorfii*, but only the large-subunit polypeptides of *N. glauca* (Kung et al., 1975) (Figure 5.2). This result indicates that *N. langsdorfii* chloroplasts were either not present or not expressed in the hybrid plant that was analyzed. The resolution of this question awaited the examination by Smith, Kao, and Combatti (1976) of additional somatic hybrids of *N. glauca* and *N. langsdorfii*. Their fusion products were similar in all respects to those isolated in the earlier experiment, except that the chromosome numbers were higher (ranging from 56 to 64), suggesting that each of the 23 independently derived hybrid plants originated from the fusion of three protoplasts. The hybridity of 16 plants regenerated from hybrid callus and of their progeny was confirmed by demonstrating that all possessed the ribulose bisphosphate carboxylase small subunits of both *N. glauca* and *N. langsdorfii*. However, only one regenerated plant contained the large subunits of both parental species. Eight plants had only the large subunit of *N. langsdorfii*, and six had only the large subunit of *N. glauca*. One plant containing exclusively the *N. glauca* large subunit and another with only the *N. langsdorfii* large subunit were regenerated from a single hybrid callus. The one hybrid plant that possessed the large subunits of both *N. glauca*

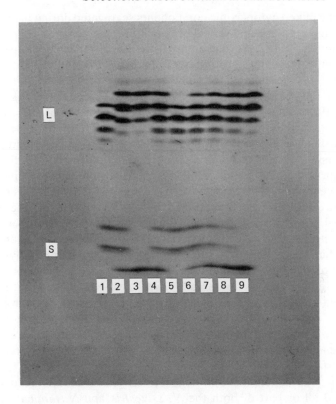

FIGURE 5.2. Polypeptide composition of ribulose bisphosphate carboxylase (RUBPCase) protein isolated from the somatic hybrid of *N. glauca* and *N. langsdorfii*. Component polypeptides of carboxymethylated RUBPCase protein were separated by isoelectric focusing in 4.5 percent polyacrylamide gel slabs containing 1 percent Ampholine, pH 5 to 7, and 8 M urea. Samples from left to right: 1, *N. langsdorfii*; 2, somatic hybrid of *N. langsdorfii* and *N. glauca*; 3, *N. glauca*; 4, a 1:1 mixture of *N. langsdorfii* and *N. glauca* proteins recrystallized as a mixture before carboxymethylation; 5, *N. langsdorfii*; 6, a 3:1 mixture of carboxymethylated proteins of *N. langsdorfii* and *N. glauca*; 7, a 1:1 mixture of carboxymethylated proteins of *N. langsdorfii* and *N. glauca*; 8, a 1:3 mixture of carboxymethylated proteins of *N. langsdorfii* and *N. glauca*; 9, *N. glauca*. L indicates large-subunit polypeptides, pH 6.5; S, small-subunit polypeptides, pH 6.0. (From Kung et al., 1975; copyright 1975 by The American Association for the Advancement of Science.)

and *N. langsdorfii* produced progeny that had only the *N. glauca* large subunit. However, plants regenerated from callus cultures that had been initiated from the leaves of this plant contained only the large subunit of *N. langsdorfii* (Chen, Wildman, and Smith, 1977). It is apparent, there-fore, that although the nuclear genomes of the hybrid plants remained stable, the mixed cytoplasmic components segregated during plant growth and during vegetative and sexual propagation.

The transfer of cytoplasmic components has also been demonstrated in the fusion of mesophyll protoplasts of *N. sylvestris* and of a male sterile plant (line 92) containing the nuclear genome of *N. tabacum* and the cytoplasm of another unknown *Nicotiana* species (Zelcer, Aviv, and Galun, 1978; Aviv et al., 1980). Because division of *N. sylvestris* protoplasts apparently is inhibited by mannitol and because line 92 protoplasts were exposed to a dose of X-irradiation sufficient to prevent their division, only hybrid fusion products were expected to form callus in a medium containing mannitol. Seven calluses were isolated, and plants were regenerated. That many of these regenerated plants were male-sterile (a characteristic conditioned by the cytoplasm of line 92) but displayed morphological features either of *N. sylvestris* alone or of both parents suggested that they were hybrids (Zelcer, Aviv, and Galun, 1978). This initial suggestion was substantiated by subsequent analysis. The progeny obtained by crossing with *N. tabacum* and *N. sylvestris* certain regenerated plants that resembled *N. sylvestris* established that regenerated plants of this type possessed only the *N. sylvestris* nuclear genome. However, crosses with regenerated plants that were morphologically similar to the sexual hybrid *N. sylvestris* × *N. tabacum* produced a heterogeneous array of progeny, some of which more closely resembled *N. sylvestris* and some of which were more like *N. tabacum*. From these results it was concluded that regenerated plants of this second class contained all or part of both the *N. sylvestris* and *N. tabacum* nuclear genomes. By means of resolving both the component polypeptides of the large subunit of ribulose bisphosphate carboxylase and the fragments produced by endonucleolytic digestion of the chloroplast DNA, it was demonstrated that these two types of regenerated plants possessed the line 92 cytoplasm. Thus hybrid plants containing the nuclear genome of one parent and the cytoplasm of the other, as well as plants containing the cytoplasm of one parent and components of both parental nuclear genomes, were recovered in these experiments. Interestingly, it appears that as in the case of the somatic hybrids of *N. glauca* and *N. langsdorfii,* regenerated plants contained chloroplasts of one or the other parent but not both parents (Aviv et al., 1980). At this point one may well begin to wonder if segregation of organelles of different origins is a general phenomenon that will prove characteristic of most somatic hybrids.

Differences in sensitivities to actinomycin D and in capacities for growth on a particular medium were exploited to isolate somatic hybrids from leaf protoplasts of *Petunia hybrida* and *P. parodii.* Because *P. hybrida* protoplasts form calluses on a medium on which protoplasts of *P. parodii* cannot undergo more than a few divisions and because division of

P. hybrida protoplasts is inhibited by a concentration of actinomycin D that does not affect *P. parodii* protoplasts, it was anticipated that only fusion products would be able to grow on medium for *P. hybrida* but supplemented with actinomycin D. Ten calluses of independent origin were recovered on the selective medium, and eight of these were induced to form plants. That these plants were somatic hybrids was suggested by chromosome counts, flower color and morphology, and electrophoretic banding patterns of leaf peroxidase isozymes (Power et al., 1976). However, genetic studies were not performed to demonstrate unequivocally that the plants were truly hybrids and not merely chimeras.

Although the hybrid genotype of the protoplast fusion products was inferred from their similarity to the sexual hybrid of *P. hybrida* × *P. parodii,* the sexual hybrid was not used in designing the selection procedure. Subsequently, more efficient selection methods were introduced by defining hormone compositions that promoted division of protoplasts of the sexual hybrid but not division of either of the parental protoplasts. Actinomycin D was included in one selective medium to eliminate a low frequency of callus formation from *P. hybrida* protoplasts (Power et al., 1977).

Selections based on visual markers

Yet another scheme devised for selecting somatic hybrids of *P. hybrida* and *P. parodii* advanced the technique to a new level; that involving the use of induced genetic markers. Protoplasts isolated from a suspension culture of an albino mutant of *P. hybrida* formed white callus on a medium on which leaf mesophyll protoplasts of *P. parodii* did not proceed beyond a few divisions; only somatic hybrids would produce green callus. Using this system, Cocking and associates (1977) obtained two tetraploid hybrid plants and one aneuploid ($4n - 2$) plant from 35 green calluses. As final confirmation of the genetic composition of two somatic hybrids, it was shown that in crosses they produced the same segregation patterns for flower colors as did two tetraploid sexual hybrids (Power et al., 1978).

The significance of using complementation of recessive albino mutations as a means of identifying somatic hybrids is that it provides a selection procedure that is more widely applicable than those based on idiosyncratic nutrient or hormonal requirements. Because albino mutations can be induced in all plant species, their use freed the design of selection systems from its former reliance on natural characteristics that could be defined only with great effort and in but few cases. This development places the investigator, rather than the plants, in control of the experiment. Instead of being able to select hybrids of only those species for which complementing resistances or growth requirements can be defined, investigators now have a technique that permits the selection of hybrids between any two species.

Melchers and Labib (1974) were the first to employ complementing albino mutations to recover hybrid plants produced by protoplast fusion. The design of this experiment (Figure 5.3) is classic and serves as a general procedure for constructing somatic hybrids. Two recessive mutants of tobacco, *subtlethal (s)* and *virescent (v)*, were used. In addition to being chlorophyll-deficient, these mutants are also light-sensitive; that is, they grow very slowly under direct illumination but grow at a nearly normal rate in weak diffuse light. Protoplasts were isolated from the leaves of haploid plants that had been obtained by culturing anthers of the homozygous diploids. Following fusion, the rare green calluses were selected, and plants were regenerated. These hybrid plants were morphologically indistinguishable from the sexually produced double heterozygotes, and approximately half had the chromosome number of the normal amphidiploid (48) (Melchers and Sacristán, 1977). An important advantage of using mutations to select hybrids is that the regenerated plants can be analyzed genetically. Recovery of both of the single homozygous recessive classes among the progeny produced by self-fertilization or by a test cross to the double mutant offers incontrovertible proof that the plant is a true hybrid and not a revertant produced by reversal of one of the original mutations. In this definitive manner Melchers and Labib (1974) selfed one of their regenerated plants and confirmed that it was indeed a hybrid by identifying both +; v and s; + individuals among the progeny.

Having used two complementing mutants of a single species to define an effective system for selecting hybrids, the next step was to employ this system to produce an interspecific somatic hybrid. The construction of putative *N. tabacum* × *N. sylvestris* hybrids was accomplished by exploiting the *sublethal* mutant of the former and a newly discovered chlorophyll-deficient mutant of the latter (Melchers, 1977). These plants, however, have not yet been characterized genetically.

FIGURE 5.3. Schematic illustration of hybridization of diploid tobacco mutants *sublethal (ss)* and *virescent (vv)* by a sexual cross (×) and of haploid mutants *(s* and *v)* by protoplast fusion (+). The nuclear genome is represented by two pairs of chromosomes, one containing the recessive *s* allele and the other the recessive *v* allele. Plastids of the two parental types are shown (open and filled symbols) to depict cytoplasmic fusion that occurs during somatic hybridization (although in this experiment the two parent cytoplasms did not differ). Haploid mutant plants grew almost normally at 800 lux, but at 10,000 lux they were pale and grew poorly, permitting selection of the hybrids. PMC, pollen mother cell; PO, germinating pollen (microgametophyte); ES, embryo sac; PL, plastids. (From Melchers and Labib, 1974.)

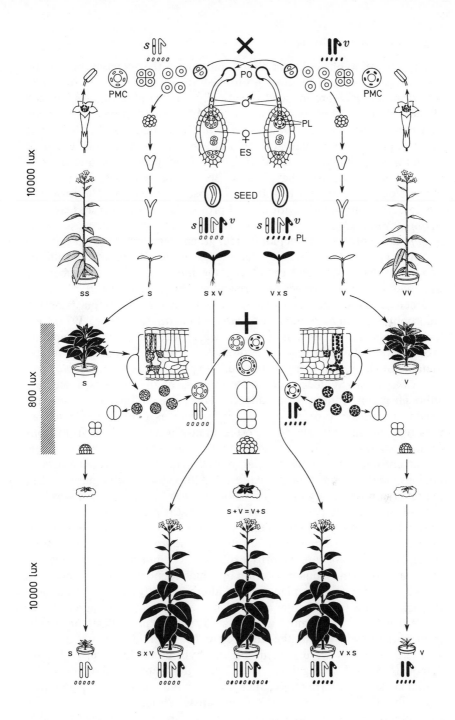

The general applicability of albino mutants to the selection of somatic hybrids is illustrated by four protoplast fusion experiments in which albino mutations were induced specifically for this purpose. An induced albino mutant of *Daucus carota* was used to construct somatic hybrids between this species and *Daucus capillifolius* (Dudits et al., 1977). But because plants regenerated at a low frequency from the normal green *D. capillifolius* protoplasts, selection for the interspecific somatic hybrids was not stringent, and they had to be distinguished on the basis of morphological characteristics and chromosome numbers. The fusion of albino carrot protoplasts with normal protoplasts isolated from leaves of *Aegopodium podagraria* (p. 99) provided a preferable system, since, as *Aegopodium* protoplasts cannot divide, green callus must have originated by protoplast fusion. Sterile plants regenerated from these calluses apparently have a normal carrot genome into which small numbers of *Aegopodium* genes have been incorporated (Dudits et al., 1979).

Green calluses were also isolated following fusion of protoplasts of two chlorophyll-deficient mutants of *Datura innoxia* (Schieder, 1977). Although calluses of various ploidies were recovered, the tetraploids produced shoots most readily. Protoplasts of one of these same albino mutants of *D. innoxia* were also fused with protoplasts of *D. discolor* and *D. stramonium* (Schieder, 1978). Protoplasts of the latter two species, although green, cannot form calluses. Therefore, following fusion with albino *D. innoxia* protoplasts, all green calluses could be selected as putative interspecific hybrids. Fertile plants were obtained from both fusion experiments. That these were hybrids was suggested by their higher chromosome numbers, chlorophyll synthesis, intermediate morphology, and isozyme banding patterns. Although in preliminary studies the reversion frequency of the two albino mutants of *D. innoxia* was less than 10^{-6} (Schieder, 1977), the possibility that the green plants could have arisen by reversion of an albino parent in the *Daucus* and *Datura* interspecific fusion experiments (or in one case from *Daucus capillifolius* protoplasts) was not satisfactorily excluded. The morphology of regenerated plants is an inadequate criterion by which to judge hybridity, because this characteristic of the plant can be altered greatly merely by passage through culture (see Chapter 3). And higher chromosome numbers can result from fusion of similar as well as dissimilar protoplasts or simply from endomitosis within the cultured cells. Definitive proof that the regenerated plants were indeed somatic hybrids could be provided by demonstration of segregation of the original parental characteristics in sexual crosses. Unfortunately, such studies have not yet been performed with either the *Daucus* or the *Datura* putative hybrids.

As chlorophyll synthesis requires both nuclear- and chloroplast-encoded functions, mutations in either of these genomes can cause chlorophyll deficiency. Thus, beyond serving simply to identify protoplast fusion products, complementing albino mutations can be used as labels of specific subcellular components. This more refined application of albino mutations was introduced by Gleba, Butenko, and Sytnik (1975) using a chloroplast mutation and the semidominant nuclear mutation *Sulfur* (*Su*) of tobacco. The *Su* mutation is normally lethal in the homozygous state (*Su*/*Su*), since plants of this genotype are yellow and cannot photosynthesize. However, homozygous *Sulfur* seedlings can be grown on a nutrient medium that provides a carbon source. Heterozygous plants are yellow green and photoautotrophic. The ability to distinguish visually *Su*/*Su* (yellow), *Su*/ + (yellow green), and +/ + (green) individuals makes the *Sulfur* mutation a convenient nuclear marker for protoplast fusion experiments. Protoplasts of the albino *Sulfur* mutant *(Su*/*Su)* containing normal plastids were fused with protoplasts isolated from albino sectors of a variegating green (+/+) plant (chloroplast mutant). Although chlorophyll synthesis and therefore complementation did not occur during growth of the hybrid calluses, yellow green plants and green plants, some of which were variegated, were regenerated. These results are what would have been predicted from the assumptions that following the initial formation of a heterokaryon the two parental nuclei were able to either fuse or segregate into separate cells (or one nucleus may degenerate) and that cytoplasmic components could also segregate independently of nuclear events (as was the case with the *N. glauca* + *N. langsdorfii* somatic hybrids) (Figure 5.4). Although nonvariegating yellow green individuals could have arisen by reversion of one *Su* allele, the fact that all of the yellow green plants that were recovered were polyploid is consistent with their having originated from fusions of the two types of parental nuclei. The possibility that variegation was due to the formation of a chimera rather than a heterogeneous cytoplasm was eliminated by demonstrating that variegated progeny are obtained from self-fertilization of regenerated variegated plants (Gleba, 1978). (If variegation is due to the chimeral association of cells containing either all normal or all mutant chloroplasts, single female gametes will carry only one type of chloroplast, and the progeny of such a plant will be nonvariegating. However, if variegation results from vegetative segregation of a mixed population of normal and mutant plastids contained within a plant cell, both types of plastid can be transmitted through a single female gamete, and variegating progeny will be produced.) Thus the various possible combinations of nuclear and cytoplasmic material that can result from protoplast fusion can be distinguished simply by visual inspection of the regenerated plants. Protoplasts of the *N. tabacum* chloroplast mutant were also fused with protoplasts of *N. debneyi* (Gleba, 1978) and of a plant that contained an *N. tabacum* nuclear genome and a male-sterile *N. debneyi* cytoplasm

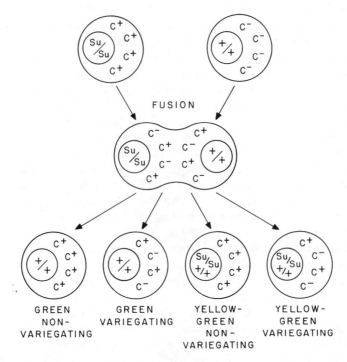

FIGURE 5.4. Diagrammatic representation of fusion of protoplasts of a homozygous *Sulfur* tobacco mutant (*Su/Su*) containing normal chloroplasts (C^+) with albino protoplasts containing a normal nuclear genotype ($+/+$) and mutant plastids (C^-). Four types of photoautotrophic products can result, depending on whether or not nuclei fuse and the parental plastids segregate.

(Gleba et al., 1978). In both cases variegation was transmitted to progeny, but the recovery of mostly diploid plants from the latter fusion indicated that karyogamy was rare. Plants regenerated from this fusion experiment were at least cytoplasmic hybrids, since the ribulose bisphosphate carboxylase large subunits were composed of both *N. tabacum* and *N. debneyi* polypeptides. Interestingly, all of these plants were male-sterile, a result that would be obtained either if male sterility is dominant or if it is determined by a cytoplasmic component other than the chloroplast (e.g., mitochondria).

Fusions between protoplasts of *N. tabacum* and the male-sterile tobacco that consists of an *N. tabacum* nuclear genome and an *N. debneyi* cytoplasm were also reported by Belliard, Pelletier, and Ferault (1977). Regenerated plants were judged to be hybrids solely on the basis of male sterility and aberrant flower morphology. Because such abnormalities appear commonly in plants regenerated from cultured cells, their initial claim of fusion could have been regarded skeptically, but comparison of the restriction endonuclease digestion patterns of the mitochondrial genomes of the two parents and of the progeny of two regenerated hybrid plants revealed that the hybrid progeny contained mitochondrial DNA fragments that were unique to each of the parents. This observation alone proved that the regenerated plants possessed a hybrid cytoplasm that must have originated from a fusion event. But even more exciting was the discovery of new DNA fragments that were not present in digests of the mitochondrial DNA of either parent. Thus, some endonuclease recognition sites were eliminated and perhaps new ones introduced in mitochondrial genomes of the somatic hybrids. These changes could result from recombination between the two parental mitochondrial genomes in the hybrid cytoplasm (Belliard, Vedel, and Pelletier, 1979). However, as other types of genetic events (e.g. deletions and duplications) could have produced the altered DNA restriction patterns, further experimentation is needed to discover their actual basis.

The *Sulfur* mutant was used also to select somatic hybrids formed by fusion of *N. tabacum* and *N. glauca* protoplasts (Evans, Wetter, and Gamborg, 1980). The inability of *N. glauca* protoplasts to divide permitted selection of hybrids as light green calluses. Regenerated plants were characterized as somatic hybrids on the basis of morphology, isozyme banding patterns, karyotype, and polypeptide composition of the ribulose bisphosphate carboxylase small subunit. In addition, regenerated plants had the light green coloration expected for individuals heterozygous for the *Su* allele, and they produced anthocyanin at the base of the leaf petiole, a feature of the *N. glauca* parent.

All of the hybrids created by protoplast fusion that we have discussed thus far could have been obtained by sexual crosses (although in the case of the interspecific *Datura* hybrids, the assistance of embryo culture is necessary). To my knowledge, in only two instances has protoplast fusion possibly accomplished the construction of a hybrid that could not be formed by other means.

In one experiment, protoplasts obtained from two chlorophyll-deficient mutants of *Datura innoxia* were fused with normal green *Atropa belladonna* protoplasts. In this case, both fusion products and *Atropa* protoplasts could form green calluses, and a marker peculiar to the *Datura* parent was needed to distinguish the hybrid. The requisite marker was provided by the pubescence of the *Datura* callus. Therefore, green pubescent calluses were selected as putative hybrids, and shoots were regenerated. But these characteristics alone could not substantiate a claim of hybridity. However, the presence in single callus and leaf cells of morphologically distinct *Atropa* and *Datura* chromosomes lends strong support to this predication (Krumbiegel and Schieder, 1979).

Somehow, experiments involving important crop species are most intriguing. And some excitement has been aroused by the reported fusion of potato protoplasts with protoplasts of a recessive yellow green tomato mutant. Although plants could not be regenerated from tomato protoplasts (and none of the regenerated plants was yellow green), no selection was operating in this experiment against the potato parent. However, regenerated plants possessed both the tomato and potato ribulose bisphosphate carboxylase small subunits. The large subunit of only a single parent was present in each of these plants, three having the large subunit of tomato and one having that of potato. The karyotypes of the regenerated plants contained a few more chromosomes than the number expected from the fusion of one tomato protoplast with one potato protoplast (Melchers, Sacristán, and Holder, 1978).Although further characterization will be required to rule out the possibility that these plants are chimeras, the initial results are most promising indeed.

Selections based on metabolic mutants

A hybrid formed from the fusion of two complementing albino mutants can be identified readily by its green coloration against a background of pale parental cells, but the ability to synthesize chlorophyll probably does not confer any growth advantage to the hybrid over the parental types. In fact, because of inherent incompatibilities, a hybrid between distantly related species may be less vigorous than the parental cells and may be overgrown. In other cases, such as that of fusion of *Sulfur* and chloroplast mutants of *N. tabacum,* complementation may not occur, and hence selection may not be applied until after plant regeneration (Gleba, Butenko, and Sytnik, 1975). The failure of albino mutations to select stringently against growth of the parental cells and in favor of the hybrids leaves us searching for other markers that will permit growth only of the fusion product. Two types of genetic markers that are employed routinely for selecting hybrids in microbial and animal cell culture systems are auxotrophic and resistance mutations. Although it has proved difficult to isolate auxotrophic mutants of plant cells, resistance mutants can be recovered fairly easily (see Chapter 4). For use in selecting hybrids, auxotrophic mutations must complement and must be recessive so that the hybrid will be prototrophic and capable of growth on an unsupplemented medium that does not support growth of either auxotrophic parent. Resistance markers must be dominant. Then the hybrid alone will grow in the presence of both antimetabolites, whereas each of the parental cell types, being resistant to only one, will succumb to the other.

As the science of plant cell genetics is still very much in its infancy, few true mutants have yet been characterized. Therefore, to date, resistance and auxotrophic markers have been exploited to select protoplast fusion products in but a small number of cases. Once more mutants have been isolated, the obvious advantages of this method for hybrid selection most certainly will compel its general adoption.

The first induced resistance marker employed to select a protoplast fusion product was provided by a kanamycin-resistant cell line of *Nicotiana sylvestris*. Chlorophyll-deficient protoplasts generated from suspension cultures of this cell line were fused with green mesophyll protoplasts of *N. knightiana*. Colonies were allowed to develop on nonselective medium and then transferred to another medium on which *N. sylvestris* colonies formed colorless callus and *N. knightiana* callus could not grow. Hybrid calluses were expected to be green. Green calluses were transferred to a medium containing kanamycin, and two calluses proved to be kanamycin-resistant. Unlike either parental cell line, these two calluses were able to produce shoots (Maliga et al., 1977). Although the calluses possessed several isozymes of both parents, definitive proof that they were hybrids has not yet been published. In these experiments, kanamycin resistance was not employed as the primary selection procedure. In contrast, following fusion of protoplasts of two complementing nitrate reductase–deficient mutants of *N. tabacum*, somatic hybrids were selected on the basis of their ability to grow on nitrate as the sole nitrogen source. The mutants used in these fusion studies were selected for resistance to chlorate (see Chapter 4) and could complement because they were defective in different components of the nitrate reductase holoenzyme (Müller and Grafe, 1978). The frequency of fusion was 30 to 50 percent, and in several experiments as many as 12% of these fusion products could be recovered as hybrid colonies (Glimelius et al., 1978). Shoots were obtained from a number of hybrid cell lines, and it is expected that the results of genetic analyses will be reported shortly.

The progression of experimental advances reviewed in this chapter has brought us ever closer to our goal of achieving facile and controlled construction of somatic hybrids (Table 5.1). Let us now consider several potential applications of protoplast fusion to plant genetics.

Marker rescue

The techniques of plant cell culture permit us to probe into a realm of genetic organization that previously was inaccessible. In conventional mutant isolation procedures the quest for abnormal plant phenotypes was confined to the recovery of morphological and color variations that did not severely affect plant maturation or viability. Furthermore, genetic analysis of such mutants demanded that they be capable of serving as at least one parent in crosses. The availability of cell culture removes many

Table 5.1. *Hybrid plants produced by protoplast fusion*

Species	Method of selecting hybrid	References
Petunia hybrida + *Petunia parodii*	Callus formation and natural drug resistance	Power et al. (1976, 1978)
	Complementation of chlorophyll deficiency and callus formation[a]	Cocking et al. (1977), Power et al. (1978)
Nicotiana tabacum + *Nicotiana tabacum*	Complementation of chlorophyll deficiencies	Melchers and Labib (1974), Gleba, Butenko, and Sytnik (1975), Kameya (1975)
	Complementation of nitrate reductase deficiencies	Glimelius et al. (1978)
Nicotiana glauca + *Nicotiana langsdorfii*	Hormone autotrophy	Carlson, Smith, and Dearing (1972), Smith, Kao, and Combatti (1976)
Nicotiana tabacum + *Nicotiana sylvestris*	Complementation of chlorophyll deficiencies	Melchers (1977)
	Complementation of damage induced by X-irradiation and callus formation[a]	Zelcer, Aviv, and Galun (1978), Aviv et al. (1980)
Nicotiana sylvestris + *Nicotiana knightiana*	Complementation of chlorophyll deficiency, callus formation,[a] and kanamycin resistance	Maliga et al. (1977)
Nicotiana tabacum + *Nicotiana debneyi*	Complementation of chlorophyll deficiencies	Gleba (1978)
Nicotiana tabacum + (*N. debneyi* × *N. tabacum*)	None	Belliard, Pelletier, and Ferault (1977)
	Complementation of chlorophyll deficiency	Gleba (1978)
Nicotiana tabacum + *Nicotiana glauca*	Complementation of chlorophyll deficiency and callus formation[a]	Evans, Wetter, and Gamborg (1980)
Daucus carota + *Daucus capillifolius*	Complementation of chlorophyll deficiency	Dudits et al. (1977)
Daucus carota + *Aegopodium podagraria*	Complementation of chlorophyll deficiency and callus formation[a]	Dudits et al. (1979)

Table 5.1. (*cont.*)

Species	Method of selecting hybrid	References
Datura innoxia + *Datura innoxia*	Complementation of chlorophyll deficiencies	Schieder (1977)
Datura innoxia + *Datura discolor*	Complementation of chlorophyll deficiencies	Schieder (1978)
Datura innoxia + *Datura stramonium*	Complementation of chlorophyll deficiencies	Schieder (1978)
Datura innoxia + *Atropa belladonna*	Complementation of chlorophyll deficiency and morphology	Krumbiegel and Schieder (1979)
Lycopersicon esculentum + *Solanum tuberosum*	Complementation of chlorophyll deficiencies	Melchers, Sacristán, and Holder (1978)

[a]Protoplasts of the hybrid and of only one parent are able to form callus on the medium employed.

of the constraints that formerly limited the spectrum of mutants that could be recovered. Because many whole-plant functions are not required for growth in culture, and because supplementation of the nutrient medium can compensate for the deletion of many otherwise essential metabolic reactions, it should now be possible to isolate cell lines that have mutant alleles of genes encoding these functions. But although many restrictions on the recovery of mutants may have been eliminated by cell culture, the genetic analysis of such mutants still can be accomplished only through sexual crosses. Therefore, an important experimental constraint that yet remains for genetic studies is that mutations cannot interfere with or preclude regeneration of a fertile adult plant.

Mutants unable to differentiate will be confined to life in a Petri dish unless their morphogenetic capacity can somehow be restored. Plant regeneration can be accomplished by suppressing, overcoming, or compensating for the interfering factor. The nitrate reductase–deficient mutants of *N. tabacum* discussed earlier illustrate this problem and its elegant solution. It was found that organogenesis could not be induced in mutant cell lines lacking detectable nitrate reductase activity. Because normal cell lines also failed to differentiate when grown on amino acids as nitrogen source, the endogenous balance of the reduced and oxidized forms of nitrogen, rather than the nitrate reductase enzyme itself, may be critical to organogenesis (Müller and Grafe, 1978). However, by means of fusing protoplasts derived from mutants lacking different activities of the nitrate reductase enzyme, hybrid cell cultures were obtained in which the two deficiencies were complemented. This hybrid cell culture possessed both nitrate reductase activity and the capacity to produce plants (Glimelius et al., 1978). Thus, in a single experiment, two mutant alleles were retrieved from cell cultures and made available for genetic analysis.

In the case of fusion of *N. knightiana* protoplasts with protoplasts obtained from a kanamycin-resistant cell line of *N. tabacum,* the somatic hybrid was able to form shoots, whereas neither parental cell type could do so (Maliga et al., 1977). Thus it is apparent that in addition to mutational interference with plant regeneration, the lack of morphogenetic capacity that may be exhibited by cells of a particular genotype cultured under certain conditions can be overcome by genetic complementation.

These observations suggest that fusion of protoplasts derived from a mutant cell line with normal protoplasts may afford a routine procedure for rescuing genetic markers that otherwise would prevent the regeneration or maturation of plants. Because the product of this fusion would be heterozygous, morphogenetic capacity would be restored only if the traits that interfere with differentiation were recessive. If it is anticipated that some mutations will obstruct organogenesis or development of a fertile plant, mutant selection experiments can be designed to avoid future headaches. Mutations can be induced in a cell line carrying an appropriate genetic marker that will facilitate subsequent selection for hybrids formed from fusion with morphogenetically competent protoplasts. A general scheme for employing protoplast fusion to rescue such mutations is illustrated in Figure 5.5. Normally the development of mature plants from auxotrophic cells would not be expected without some means of providing the required supplement during ontogenesis. However, auxotrophic mutants can be sought in a cell line carrying a dominant allele that confers resistance to a given antimetabolite. Following fusion of resistant auxotrophic protoplasts to normal protoplasts, the hybrid, but neither of the parents, will be capable of growth on a minimal medium containing the antimetabolite. To eliminate selection for prototrophic revertants of the resistant auxotrophic cell line, a dominant allele for resistance to a second compound should be incorporated into the other parental cell line. Then only the hybrid possessing the two resistance markers will divide in the presence of both antimetabolities. Recovery of other undesirable types arising from spontaneous mutation of one parent to resistance to the other antimetabolite can be avoided by inclusion of additional genetic markers. Complementing auxotrophic mutants can also be used as the

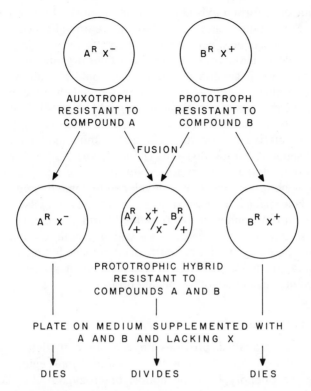

FIGURE 5.5. Scheme for marker rescue by protoplast fusion. An auxotrophic mutation (requiring compound X for growth) is induced in a cell line resistant to antimetabolite A. Protoplasts of the auxotroph are fused with protoplasts of a prototrophic cell line resistant to antimetabolite B. Only the hybrid will be capable of growth on a medium lacking X and supplemented with A and B.

two parental cell lines; in this case only the prototrophic hybrid will grow on a minimal medium. Either method will select for a fusion product that harbors the new allele in a heterozygous state. As long as this allele is recessive, mature fertile plants can be regenerated from the hybrid cells. Thus the new allele will have been recovered from the cultured cells and placed in a form that will make conventional genetic analysis possible.

Toward a parasexual cycle

Traditional genetic analysis of plants is a procedure that is protracted by its reliance on crosses with whole organisms that have rather long life cycles. Nevertheless, this procedure must be endured to accomplish genetic recombination by the alternation of karyogamy (the fusion of two gametic nuclei to form a hybrid zygote) and meiosis. In this manner, two sets of genes from different lines of descent are brought together and then segregated in subsequent generations. The patterns in which these genes segregate yield valuable information about their organization, regulation, and function. But if genetic recombination could be achieved by other means, we would be delivered from our present dependence on the sexual cycle, and the process of genetic analysis might be greatly accelerated. In fungi and cultured animal cells a parasexual cycle is implemented by fusion of vegetative or somatic cells and elimination of chromosomes during subsequent mitotic divisions (Pontecorvo, 1958). Now it appears that the components necessary for establishment of a parasexual genetic system in higher plants have been developed to some degree: protoplasts of many plant species have been fused, and in several cases it has been shown that chromosomes are lost from the hybrid genome. Perhaps, then, a brief description of the methods by which genetic complementation, linkage, and mapping studies are performed in some fungi and animal cell cultures is appropriate, as these techniques might prove applicable to cultured plant cells in the not too distant future.

Complementation tests

Complementation analysis provides a means of rapidly determining if two recessive mutations are located in the same functional region (cistron) by examining the phenotype of the double heterozygote. In plants, complementation tests are usually performed by crossing two homozygous mutants, but in microbial and animal cell systems the hybrids are produced simply by cell fusion. Complete restoration of a normal phenotype indicates that the mutations are in separate cistrons (intercistronic complementation). In cases in which an enzyme is formed by aggregation of the polypeptide products of a single gene, it is possible for the differently defective polypeptides encoded by distinct alleles of that gene to form a partially functional enzyme molecule. Allelic or intracistronic complementation typically will effect incomplete restoration of the normal phenotype.

By now it is evident that the double heterozygotes prerequisite for complementation analysis can be produced by fusing protoplasts isolated from two mutant cell lines. Allelism of two mutations then might be determined without first proceeding through plant regeneration and maturation. This method was used to investigate the genetic relationship of nitrate reductase–deficient cell lines of *N. tabacum*. The presence of nitrate reductase activity in the somatic hybrids and their ability to grow on nitrate demonstrated that the two mutations were located in different genes (Müller and Grafe, 1978; Glimelius et al., 1978).

Linkage determination

The analysis of genetic linkage in somatic hybrids exploits the random loss of chromosomes that occurs during cell multiplication. In *Aspergillus*, haploids are produced spontaneously from vegetative diploid cells at a low frequency (Pontecorvo and Käfer, 1958) that can be increased by treatment with *p*-fluorophenylalanine (Lhoas, 1961). Haploidization yields all possible recombinations between chromosomes, but practically none between markers on the same chromosome. Thus, if in all haploid clones examined the presence or absence of two markers is correlated, the genes have been shown to segregate as a unit and be linked. The genes are unlinked if in a fair proportion of the clones one marker appears without the other.

Human chromosomes are shed preferentially from human + rodent hybrids, but this loss is more gradual and haphazard than in the *Aspergillus* system. Nevertheless, the fact that the assortment of human chromosomes retained varies from one clone to another permits genetic linkage relationships to be determined. Linkage is inferred from the simultaneous loss or occurrence of two markers in all clones, and lack of linkage is inferred from their segregation (Figure 5.6). By the use of selective markers, experiments can be designed to retain specific chromosomes. For example, following fusion of rodent cells deficient for hypoxanthine-guanine phosphoribosyl transferase (HGPRT) with human cells, only clones retaining the human chromosome that carries the intact homologous human gene will survive on a medium demanding HGPRT function. Linkage to the HGPRT locus is then established for additional markers appearing in clones that contain only the single human chromosome. The particular human chromosome associated with the linkage group can then be identified by analysis of fluorescent banding patterns.

Genetic mapping

A chromosome map is usually constructed on the basis of meiotic recombination frequencies and describes the linear sequence of genes and the relative distances between genes on an individual chromosome. Because this procedure is not applicable to animal cell cultures, other techniques to establish gene order have been devised.

One method of localizing genes on the chromosome exploits rearrangements of genes that result from chromosome translocations. Translocations are structural changes in which a fragment of one chromosome is transferred to another chromosome. As a consequence of such an event, genes that have been unlinked can become associated on the same linkage group. In one experiment, human cells carrying a translocation

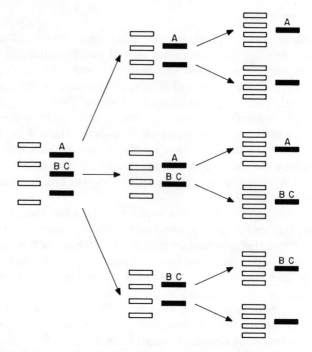

FIGURE 5.6. Diagram of clones segregating during successive mitoses of human + rodent hybrids containing three human chromosomes, one of which carries marker A and one of which carriers linked markers B and C. Rodent chromosomes are illustrated as open bars that are retained through all nuclear divisions. Filled bars represent human chromosomes that are eliminated during nuclear divisions.

that caused transfer to an autosome of the segment of the human X chromosome containing the HGPRT locus were fused with HGPRT-deficient rodent cells. Hybrids retaining the translocated chromosome were then selected as the remainder of the human chromosomes were eliminated during successive nuclear divisions. When these clones were examined, it was discovered that a previously unlinked autosomal marker (nucleoside phosphorylase) was segregating with the HGPRT locus. This marker could then be assigned to the autosome that was involved in the translocation (Ricciuti and Ruddle, 1973). By determining whether or not previous linkage relationships have been disrupted, one can use this same method to determine if two markers are located on the same segment of a given chromosome. If, for example, the translocation that we have just considered separated other X-linked markers from HGPRT, it would be apparent that these markers were not in the region of the X chromosome that was transferred to the autosome. However, for detailed mapping, this technique would require a series of defined translocations with different breakpoints that would serve to subdivide each of the chromosomes. The unavailability of such translocation series severely limits the utility of chromosome rearrangements for genetic mapping.

In another method for mapping genes, human cells are first irradiated to induce breaks in the chromosomes. Following fusion with rodent cells lacking HGPRT activity, hybrids containing the functional human gene are selected. But since the human chromosomes have been fragmented, genes that are normally linked may have been separated. Because the probability of two linked human genes being included in a single segment is inversely related to the distance between them on the chromosome, the order of genes on the chromosome and the relative distances between them can be established by the frequencies with which unselected human markers are cotransferred with the HGPRT locus. Thus the unselected marker that appears in the greatest number of HGPRT-containing hybrid clones must be most closely linked, and the marker appearing in the smallest number of clones must be located farthest from the HGPRT locus on the chromosome (Goss and Harris, 1975).

In fungal systems somatic recombination is used routinely for mapping genes on chromosomes (Pontecorvo and Käfer, 1958; Manney and Mortimer, 1964).To perform this method of analysis, one must begin with diploid vegetative cells that are heterozygous for the desired markers. Because mitotic recombination is a rare event (although the frequency can be increased by treatment with chemicals or X-rays), markers that permit selection of the recombinant types are required. Now, if during mitosis of the heterozygous diploid nucleus a crossover occurs between two chromatids of the homologous chromosomes, one of the two possible segregation patterns will yield daughter nuclei homozygous for the chromosome segment distal to the point of the crossover (Figure 5.7). If the marker is located in the distal chromosome segment the event will be detectable phenotypically by the appearance of a b/b clone and a $+/+$ clone (twin spot). Because the probability of a crossover occurring in the region between the marker and the centromere is directly related to the length of that region, the frequency with which homozygous clones appear can be used as a measure of map distance. Thus the higher the frequency of mitotic recombination the farther the marker from the centromere. Mapping by mitotic recombination also permits unambiguous assignment of the order of genes on the chromosome. All clones homozygous for the more proximal marker (a in Figure 5.7) will also be homozygous for b (crossover in the region between a and the centromere), but not all b/b clones will be homozygous for a (crossover in the region between a and b).

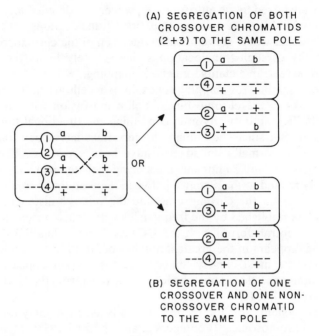

FIGURE 5.7. Diagram showing mitotic recombination and its genetic consequences in a doubly heterozygous chromosome pair. In segregation A, daughter
nuclei remain heterozygous, and no visible change results. Segregation B produces daughter nuclei homozygous for the segment distal to the point of exchange.
(Adapted from Pontecorvo, 1958.)

The possibility of applying mitotic recombination to genetic mapping in
higher plants was first explored by Carlson (1974) using *N. tabacum*.
Plants heterozygous for the semidominant *Sulfur* mutation (*Su*) and for
the recessive *chimeral(cl)* mutation were constructed with the markers in
repulsion:

$$\frac{cl\ +}{+\ Su}$$

Tissue heterozygous at the *Sulfur* locus is light green and can be
distinguished readily from the albino homozygous tissue and the green
normal tissue. *Chimeral (cl)* causes chlorotic blotching of the leaf tissue
and is located 38 map units from *Sulfur* on the same chromosome.
Crossover events in the region between the centromere and the *Su* locus
can be recognized as giving rise to albino and green twin spots on the
background of light green heterozygous tissue (Figure 5.8). Such symmet-
rical sectors of homozygous tissue cannot be produced simply by rever-

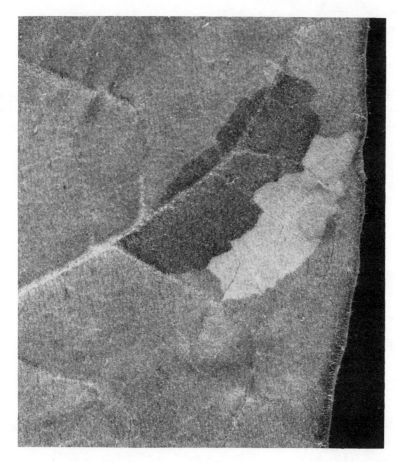

FIGURE 5.8. Yellow (light) (*Su/Su*) and green (dark) (+/+) twin spots on the leaf of a *Su, +/+, cl* tobacco plant. (From Carlson, 1974.)

sion. Carlson (1974) was able to identify twin spots in leaf tissue and in callus cultures and regenerate plants from the green sectors. By means of crosses with homozygous *chimeral (cl, +/cl, +)* plants, all 11 plants that had been regenerated from independently occurring green sectors were shown to be heterozygous for *chimeral* (the progeny were half normal and half *chimeral*) and therefore had the genotype

$$\frac{cl \quad +}{+ \quad +}$$

If the *chimeral* locus were distal to *Sulfur* and on the same chromosome arm, all plants homozygous for the normal allele of the *Sulfur* locus would also be homozygous *chimeral (cl, +/cl, +)*. The observed results are consistent with *chimeral* being either proximal to *Sulfur* or on the opposite arm of the same chromosome.

Carlson's experiment was the first realization of our aspiration to perform genetic analysis directly with cultured plant cells. By means of determining the frequency of green clones in a population, the *Sulfur* locus could be mapped relative to the centromere. We can be confident that the use and sophistication of parasexual genetics will increase as more markers that are expressed in vitro are characterized.

6

Organelle uptake

Released from behind the barrier of the cell wall, the plant protoplast is able to engage in a variety of enterprises from which the intact cell is excluded. In addition to fusion (see Chapter 5), protoplasts are capable of incorporating large particles such as macromolecules, bacteria, viruses, and organelles. In the previous chapter we considered fusion of protoplasts as a means of creating new combinations of genetic material. As we have seen, protoplast fusion can bring together all of the components of two plant cells of very different origins. Now we shall investigate transplantation of specific organelles into protoplasts as a more selective approach to producing a subcellular potpourri.

By introducing a single class of organelles into a foreign protoplast, one can study more accurately the relationships among mitochondria, chloroplasts, and the nucleus. In particular, this experimental approach may prove valuable in exploring the degree of genetic autonomy of chloroplasts and mitochondria and the extent and nature of their dependence on the nucleus. Organelle uptake will also allow important plant traits that are determined by chloroplast and mitochondrial genomes to be selectively transferred between species. In the more distant future, chloroplasts and mitochondria may become useful as vehicles for transferring foreign genes into plants. This exciting possibility is invited by their relatively small DNA content (compared with that of the nucleus) and the ease with which they can be isolated and reintroduced into protoplasts.

The incorporation of organelles by protoplasts is accomplished by a rather straightforward procedure in which protoplasts are incubated with a preparation of isolated mitochondria, chloroplasts, or nuclei. Various agents, such as polyethylene glycol, may be added to increase the frequency of incorporation. But interpreting the results of an organelle uptake experiment is more difficult.

Cytological determination of uptake

If uptake of organelles is monitored by light microscopy, it is difficult to distinguish between mere adsorption of the organelle to the plasmalemma

125

and actual incorporation of the organelle into the cytoplasm. When viewed through the microscope, an organelle adhering to the plasmalemma will appear to be inside a protoplast, the orientation of which places the organelle either above or beneath the equatorial plane of the protoplast. However, if the preparation is rolled between the slide and the coverslip to position the organelle in the equatorial plane, an organelle adsorbed to the membrane will appear outside the protoplast, whereas one that has been incorporated will continue to appear inside (Potrykus and Hoffmann, 1973). If the organelle is located in the apical or basal region of the protoplast, the side of the membrane on which it lies may be determined by through-focusing. Additional cytological evidence of incorporation is sometimes offered by visible effects of the organelle on the internal structure of the cytoplasm. Bonnett and Eriksson (1974) confirmed uptake of a *Vaucheria* chloroplast into a carrot protoplast by observing that the tonoplast was indented as a result of a chloroplast wedged between the plasmalemma and the vacuole (Figure 6.1).

In distinction to direct uptake of an organelle by pinocytosis following mixing of protoplasts and organelles is the possibility of fusion of the outer membrane of an organelle with the plasmalemma of the protoplast. In such a case the unruptured inner membrane and its contents may enter the cytoplasm of the protoplast or may remain associated with the plasmalemma. Viewed through the light microscope it might appear that uptake has occurred in either circumstance, but because it is unlikely that a new organelle outer membrane will be regenerated, neither event can be said sensu stricto to represent incorporation of an intact organelle (Davey, Frearson, and Power, 1976).

It is apparent, therefore, that conclusive cytological evidence of organelle uptake is difficult to obtain. The laboriousness of this procedure also limits the number of cells that can be examined. The most serious disadvantage of the cytological method is that it sacrifices the specimen and therefore the possiblity of inquiring into the most interesting questions of expression and replication of the organelle within the recipient protoplast.

Organelle function as an assay of uptake

In appropriate cases the most convincing evidence for incorporation (and the most informative) is the demonstration of function of the foreign organelle in its new environment. If incorporation of a functional organelle proves to be a rare event, obtaining expression of the transplanted organelle may offer the additional benefit of permitting selective procedures that enable large numbers of protoplasts to be screened. However,

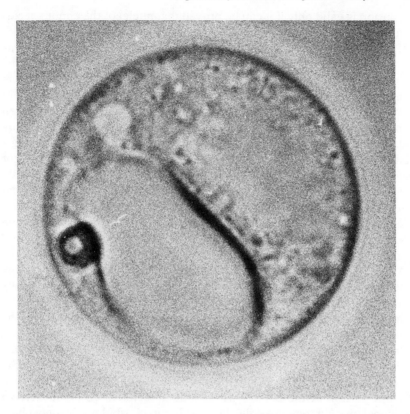

FIGURE 6.1. Incorporation of a *Vaucheria* chloroplast into a protoplast of *Daucus carota*. The position of the chloroplast between the vacuole and the plasmalemma has caused an indentation of the tonoplast. (From Bonnett and Eriksson, 1974.)

experiments of this nature must be designed properly to guard against three possible sources of spurious results.

First there is the technical problem of contamination of the organelle preparation (e.g., chloroplasts) with other organelles (e.g., nuclei). In certain cases one may be led to believe that a given result is the consequence of the introduction of the single organelle, whereas in fact the presence of a second component is also required to produce the observed effect. For example, following the introduction of chloroplasts of one species into protoplasts of another, the replication and expression of these donated chloroplasts might indicate that they are not dependent exclusively on a nucleus of the same species but that the nucleus and cytoplasm of the host species can support their functioning. But this conclusion is valid only if it can be proved that nuclei of the donor species were not introduced inadvertently together with the chloroplasts.

Uncritical reliance on genetic markers to detect organelle uptake opens the door to the second danger: the problem of contamination of the recipient cell with organelles of the donor phenotype. This problem commonly arises in experiments in which normal green chloroplasts are introduced into protoplasts obtained from albino sectors of a variegated mutant plant. A variegated plant is composed of some cells with only mutant plastids, some having only normal plastids, and some containing a mixture of both types. Even if protoplasts are isolated from a leaf sector that appears to be completely albino, it is possible that a few protoplasts will contain a small number of normal chloroplasts. Preferential replication of these normal chloroplasts will produce a green cell. Therefore it is necessary to furnish additional evidence that greening results from incorporation of a foreign chloroplast rather than from multiplication of a normal chloroplast deriving from the variegated plant.

A third source of misinterpretation, also intrinsic to experiments using genetic markers, lies in the possibility that restoration of organelle function may occur by reversion of the mutation in the recipient system rather than by uptake of an intact organelle. For example, in the chloroplast uptake experiment described earlier, if the chlorophyll deficiency of the recipient cell is due to a point mutation of the chloroplast genome, reversion of that mutation can give rise to a colony of green cells. Such an event can be excluded by the use of nonrevertible mutants in which a portion of the chloroplast genome is deleted. Alternatively, an additional marker (such as the large subunit of ribulose bisphosphate carboxylase) can serve to distinguish the introduced chloroplast from those of the recipient line.

Uptake of chloroplasts

The visibility of pigmented chloroplasts has made them the favored material for studies of organelle uptake. Incorporation of green chloroplasts can be monitored easily in albino protoplasts obtained from mutants or from plant cells grown under conditions that prevent chlorophyll synthesis. Using a model system devised for studying chloroplast uptake, Bonnett and Eriksson (1974) followed the incorporation of chloroplasts of the green alga *Vaucheria* into protoplasts of carrot. The design of this system permitted them to conclude with fair certainty that the presence of a green organelle in a protoplast resulted from uptake of an algal chloroplast. The carrot protoplasts were obtained from cell suspension cultures and contained only leukoplasts and no mature chloroplasts. Furthermore, under the experimental conditions employed, development of carrot chloroplasts did not occur. Identification of the green algal chloroplasts was unambiguous because of their distinctive size and ultrastructure. On the basis of microscopic examination following incubation of a mixed suspension in the presence of polyethylene glycol, it appeared that algal chloroplasts had been incorporated by as many as 16 percent of the carrot protoplasts.

A cytological study of uptake of higher plant chloroplasts was performed by Potrykus (1973), who used normal *Petunia hybrida* chloroplasts and albino protoplasts isolated from wholly white leaves of a variegated mutant *P. hybrida* plant. Maternal inheritance of the chlorophyll deficiency suggested that loss of chloroplast function was due to a mutation in the chloroplast rather than in the nuclear DNA. Since the recipient protoplasts contained intact nuclei, any normal chloroplasts that had been incorporated should have been able to function. When a mixed suspension of chloroplasts and protoplasts was centrifuged at low speed, green chloroplasts were incorporated into as many as 0.5 percent of the protoplasts. However plants were not regenerated, and convincing evidence that the green chloroplasts had originated from outside and afterward were actually contained inside the protoplasts was lacking.

Although the chlorophyll content of chloroplasts may serve adequately as a marker for the initial monitoring of uptake, it is by no means an infallible indication of incorporation as it is not an exclusive characteristic of the introduced chloroplast. A condition necessary to demonstrate uptake unambiguously is that the introduced chloroplasts possess a distinctive feature that cannot be manifested by the plastids of the recipient protoplast. An unequivocal marker of this type is found in the species-specific isoelectric focusing pattern of the component polypeptides of the chloroplast-encoded large subunit of ribulose bisphosphate carboxylase (RUBPCase) (see Chapter 5). The presence of polypeptides of both species in the photosynthetic products formed by mixing chloroplasts of one species with protoplasts of another provides proof of uptake. Such a demonstration was achieved by Carlson (1973b) with chloroplasts of *Nicotiana suaveolens* and protoplasts obtained from white leaf tissue of a variegating albino mutant of *N. tabacum*. In contrast with previous experiments in which chloroplast uptake was monitored microscopically, in this experiment uptake was scored on the basis of green callus formation among the regenerating protoplasts. By this criterion chloroplast incorporation occurred at a frequency of 2×10^{-4}, which greatly exceeded the reversion frequency of approximately 10^{-8} at which green colonies appeared in the absence of added chloroplasts. Efforts to regenerate plants from the green calluses yielded only a single individual. This plant was variegating and sterile and had a chromosome number of 64 (*N. suaveolens* and *N. tabacum* have 32 and 48 chromosomes, respectively). Polypeptides composing the RUBPCase large subunits of both *N. sauveolens* and *N. tabacum* were present in the regenerated plant. This result indicated that the plant contained chloroplast DNA of both species. Quite unexpected, however, was the discovery in the leaf extracts of the component polypeptides of the RUBPCase small subunits of the two species. It was evident that nuclear DNA from *N. sauveolens* had also been incorporated into the *N. tabacum* protoplast (Kung et al., 1975). Since the one regenerated plant contained both chloroplast and nuclear genes of *N. suaveolens,* this experiment answered one question, but left another to tantalize us. The results clearly demonstrate that isolated chloroplasts can be transferred into protoplasts. However, although the introduced *N. suaveolens* chloroplasts can function in the *N. tabacum* cytoplasm, it is not known if nuclear genes of the same species must also be transferred for them to do so. The specificity of nuclear–organelle interactions will be discerned only by examination of greater numbers of heteroplasmic plants obtained by additional experiments.

Uptake of nuclei

The evolution of nuclear uptake experiments has paralleled that of studies of chloroplast uptake. These experiments progressed from scrutiny under the light microscope to a search for expression by protoplasts of genetic markers contained in the isolated nuclei.

In the first nuclear uptake experiment, Potrykus and Hoffmann (1973) were able to observe incorporation of *Petunia hybrida* nuclei that had been stained with ethidium bromide into protoplasts of *P. hybrida, Nicotiana glauca,* and *Zea mays.* Fluorescence of the stained nuclei under ultraviolet light made them readily distinguishable from nuclei of the recipient protoplasts. When alternating layers of protoplasts and isolated nuclei were centrifuged in a hypotonic medium containing lysozyme, stained nuclei were incorporated into 0.5 percent of the protoplasts. Subsequently the uptake efficiency was improved by use of a different technique. Uptake of nuclei (of rye, barley, or wheat) by 3 to 5 percent of a population of maize protoplasts was accomplished by adding polyethylene glycol to a mixed suspension of nuclei and protoplasts on a coverslip (Lörz and Potrykus, 1978). But, as in the case of chloroplast uptake, the sole and less than satisfactory assay of incorporation was whether or not a nucleus continued to appear inside a protoplast that was rolled this way and that.

In an attempt to provide genetic evidence of nuclear transplantation, Lörz and Potrykus (1978) employed the complementing recessive nuclear mutations of tobacco, *sublethal* and *virescent,* which had been used to identify somatic hybrids produced by protoplast fusion (see Chapter 5,

Figure 5.3). After mixing *virescent* protoplasts with *sublethal* nuclei and vice versa, 1.8 × 10⁷ calluses were isolated and incubated under high light intensity. Under these conditions, only complementing heterozygotes, not the parental mutant types, should survive. Although 40 green calluses were recovered from an experiment in which *sublethal* protoplasts were combined with *virescent* nuclei in the presence of polyethylene glycol, all regenerated plants were *sublethal*. Now one is left pondering the failure of such a well-designed and well-executed experiment to yield a positive result. One possibility is that no nuclei were incorporated into the protoplasts. Another explanation assumes that nuclei were introduced but that they were so extensively damaged by the isolation and incubation procedures that they were unable to function once they finally gained entrance to the new cytoplasm.

The use of nuclei enclosed in membrane vesicles may increase the frequency of uptake, minimize the trauma of isolation, and protect nuclei from injury during the period of incubation with protoplasts. Membrane-bound organelles (subprotoplasts) can be isolated as cytoplasmic buds formed by protoplasts and as the products of protoplast lysis (Binding and Kollmann, 1976). Subprotoplasts that contain only a nucleus have been produced by centrifuging protoplasts in the presence of cytochalasin B (Wallin, Glimelius, and Eriksson, 1978). The contents of subprotoplasts can then be introduced into intact protoplasts simply by fusion (Binding and Kollmann, 1976). Perhaps in the near future subprotoplasts will also prove useful as vehicles for introducing isolated DNA fragments or chromosomes into plant protoplasts.

7

Introduction of purified DNA

The induction and selection of mutations among cultured cells offer a most valuable means of exploring and modifying the genetic organization of higher plants (see Chapter 4). Yet this technique, at present still far from fully explored, is ultimately limited by its dependence on the resources of the plant genome. One can only hope to recover novel phenotypes that result from small changes in nucleotide sequences or structural rearrangements of existing genetic material. Other methods of altering the plant genotype that do not suffer these constraints involve the introduction of foreign genetic information either by protoplast fusion (see Chapter 5) or by the uptake of organelles (see Chapter 6). Now one no longer need trust to the chances of mutation, but instead can think in terms of incorporating into a plant genome known genes from other organisms. But even these techniques, as fantastic in promise as they may seem, are far from ideal.

We must consider that any single gene is present at but a very low concentration in the donor nuclear or other organelle genome and that the transfer and stable incorporation of DNA can be expected, at best, to occur at a low frequency. Thus, isolation of those rare recipient cells containing the desired DNA sequence could prove arduous or, if no means of selection is available, impossible. In addition, it is quite probable that any cell that has acquired the desired genetic information will also have incorporated undesirable or deleterious sequences. Moreover, simply transferring genes via protoplast fusion or organelle uptake does not take advantage of the possibility of modifying the donor DNA by any of a number of recently developed chemical or enzymatic procedures. That is, the donor DNA sequence is introduced into the recipient cell largely in its native form and cannot be tailored to better suit the unaccustomed demands of its new environment and tasks.

However, if DNA is removed from the donor cell or organelle before being introduced to the recipient cell, so too are removed the experimental obstacles posed by the packaging of DNA in these cumbersome and inaccessible forms. And when one is not limited to the use of plant protoplasts or organelles as vehicles for introducing foreign DNA, the sources from which that DNA can be obtained also become unlimited.

132

The use of isolated DNA makes it possible to exploit recombinant DNA technology to purify specific DNA fragments by cloning them in bacteriophages or plasmids *(vide infra)*. In addition, one can attempt to introduce genetic information that previously was unknown in the plant kingdom. It is, for example, by this method that many researchers hope to provide plants with the capacity to reduce molecular nitrogen to ammonium, a capacity now possessed exclusively by prokaryotic species. However, as will be discussed later, problems also arise in introducing naked DNA into cultured cells or protoplasts.

But before proceeding we must pause to consider a controversial matter of terminology. The term "transformation" was introduced to describe the transfer of genetic information in bacterial systems by means of exogenously supplied naked DNA. "Transduction" refers to the virus-mediated transfer of genes between cells. Since the introduction of these terms, some understanding has been attained of the molecular mechanisms involved in the actual processes. Consequently, the terms "transformation" and "transduction" are understood by many to specify not only the transfer of markers by different means of conveyance but also the mechanisms by which uptake and integration of the donor DNA occur.

When Doy, Gresshoff, and Rolfe (1973a,b) reported the bacteriophage-mediated transfer and expression of bacterial genes in cultured plant cells, they found it necessary, because of their ignorance of the mechanism by which this phenomenon was accomplished, to describe their observations with a new term devised to avoid any assumption of molecular mechanisms implicit in bacterial terminology. Accordingly, these authors introduced the term "transgenosis" to describe "the transfer of genetic information from one cell to another, followed by phenotypic expression." Transgenosis could refer to the introduction of foreign genes as isolated nucleic acids or through an intact virus, organelle, or protoplast. The need for this term, however, is not entirely clear. Exogenous DNA is not incorporated into *Escherichia coli* and *Bacillus subtilis* or *Haemophilus influenzae* by the same mechanism; yet the process of uptake and expression of donor DNA by all of these organisms is referred to as transformation. Similarly, the term transformation is used to describe the incorporation and expression of exogenous DNA by yeast (Hinnen, Hicks, and Fink, 1978), *Neurospora* (Case et al., 1979), and mammalian cells (Wigler et al., 1979). And the transfer of bacterial genes to mammalian cells by means of a bacteriophage vector has been referred to as transduction (Merril, Geiser, and Trigg, 1974). Hence it becomes difficult to justify objections to denoting the introduction of foreign genes into plant cells as naked DNA or by means of a phage vector as transformation and transduction, respectively. I do not desire to enter into this dispute but rather, by presenting the problem, to accelerate its resolution. Accordingly, in the ensuing discussion all these terms have been eschewed in favor of more descriptive, albeit more cumbersome, phraseology.

Uptake of isolated DNA

Several events are required for genetic information to be introduced in the form of isolated DNA and then expressed by a plant cell: that the exogenous DNA be taken up as an intact macromolecule, that it enter an organelle (nucleus, chloroplast, or mitochondrion) possessing an apparatus capable of accomplishing its replication and transcription, that the transcript be correctly processed and translated, and that the protein product be able to function in its new molecular environment. Integration of the foreign DNA into the host genome might facilitate, but not necessarily be required for, its replication.

The first barrier to the penetration of exogenous DNA is the cell wall. Heyn and Schilperoort (1973) found that following incubation of tobacco cells in a solution of *Agrobacterium tumefaciens* DNA, only 0.145 percent of the bacterial DNA associated with the cells remained associated with protoplasts prepared from those cells. But even protoplasts possess formidable defenses against the intruding DNA (Figure 7.1). Exogenous DNA may be adsorbed to the plasmalemma or degraded by extracellular nucleases. Molecules that do enter the protoplast must survive encounters with nucleases in the cytoplasm before confronting the organelle membrane. Two different approaches might be followed to detect the uptake and incorporation of exogenous DNA by a cell. A geneticist would examine the recipient cell population for individuals that express a trait encoded by the donor DNA. Given effective selection procedures for the trait, this approach is very sensitive and theoretically could detect the expression of foreign genetic information by a single member of an infinitely large population. However, such expression would require that the entire series of steps of uptake, replication, transcription, and translation be accomplished. Failure to observe expression might result from inability to perform any one of these distinct functions (e.g., initiation of RNA or protein synthesis) and therefore would not afford information about the achievement of these steps individually. Alternatively, a biochemical approach, although less sensitive, at least would reveal whether or not the exogenous DNA is being taken up and, if so, to what extent it is being degraded. Because a detailed description of such experiments is not to the point of this discussion and, in any case, is provided by several recent reviews (Chaleff and Polacco, 1977; Kleinhofs and Behki, 1977; Lurquin, 1977; Ohyama, Pelcher, and Schaeffer, 1978),

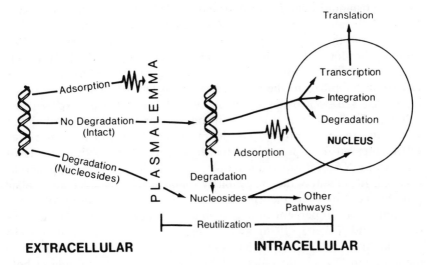

FIGURE 7.1. Schematic representation of possible fates of exogenous DNA after its introduction to plant protoplasts. The same scheme might be depicted for uptake of DNA by chloroplasts or mitochondria, except that in such cases translation would occur within the organelle rather than in the cytoplasm as shown.

no more than a brief summary of the general procedure and results is offered here.

In experiments designed to detect DNA uptake biochemically, protoplasts first are incubated in a solution of radioactively labeled DNA. Next, the protoplasts are washed with unlabeled DNA or treated with deoxyribonuclease to remove labeled DNA that is adsorbed to the plasmalemma but that has not entered the protoplast. Finally, the amount of exogenous DNA that has been incorporated is determined by measuring the radioactivity associated with the washed protoplasts or with the DNA that is extracted from these protoplasts. In some cases the protoplasts are fractionated to determine the amount of radioactivity associated with specific subcellular components. In experiments in which the introduced DNA has a different buoyant density than that of the recipient cell, the extent to which the foreign DNA has been degraded can be ascertained by density gradient centrifugation of DNA released by lysing the protoplasts. If extensive degradation of the exogenous DNA has occurred, most of the radioactivity will be associated with the band appearing at the buoyant density of the recipient cell DNA. But if the introduced DNA is largely intact, a radioactively labeled band will appear at the buoyant density characteristic of the foreign DNA, in addition to an unlabeled band of recipient cell DNA. A disadvantage of this procedure is that a substantial quantity of exogenous DNA must be incorporated to form a detectable band. A far more sensitive means of identifying donor sequences in the recipient protoplasts is by DNA-DNA filter hybridization. If the donor DNA has been completely degraded, the recipient protoplasts will not contain donor-specific sequences. In this case, donor DNA will hybridize to the same extent with DNA isolated from protoplasts that had been exposed and those that had not been exposed to a solution of donor DNA. However, if some donor sequences remain intact within the recipient protoplasts, more donor DNA will hybridize with DNA prepared from protoplasts that had been treated with donor DNA than with DNA of untreated protoplasts. In such experiments, generally less than 10 percent of the exogenously supplied DNA is taken up by plant protoplasts. However, incorporation by *Nicotiana tabacum* protoplasts of as much as 30 percent of added single-strand bacteriophage fd DNA (Suzuki and Takebe, 1976) and double-strand bacteriophage λ DNA (Suzuki and Takebe, 1978) has been reported. The extent to which incorporated DNA is degraded will vary according to the origin of both the donor DNA and the recipient protoplast. It was found that very little intact *E. coli* DNA could be recovered from *Nicotiana glutinosa* (Uchimiya and Murashige, 1977) or *Ammi visnaga* (Ohyama, Pelcher, and Schaeffer, 1978) protoplasts. However, approximately 25 percent of λ DNA (Suzuki and Takebe, 1978) and 30 percent of fd DNA (Suzuki and Takebe, 1976) taken up by *N. tabacum* protoplasts still retained its initial size after an incubation period of 60 minutes. A large proportion of *E. coli* plasmid DNA was recovered as fragments of at least 2×10^6 d 24 and 72 hours after its uptake by barley and tobacco protoplasts, respectively (Hughes, White, and Smith, 1979). Following incorporation by cowpea mesophyll protoplasts, this same plasmid (pBR313) was cleaved into two or three segments, whereas 21.5 hours after its incorporation by turnip mesophyll protoplasts, no significant degradation could be detected (Fernandez, Lurquin, and Kado, 1978). The discovery that a considerable portion of this intact donor DNA is associated with the nucleus (Lurquin and Kado, 1977; Suzuki and Takebe, 1978; Hughes, White, and Smith, 1977, 1978) offers encouragement that this method may be useful for introducing genetic information into plants.

As an alternative to radioactive isotopes, genetic markers may be used to label DNA that is being introduced into plant protoplasts. Incorporation of exogenous DNA then can be detected as expression of the marker by the recipient cell. Use of donor DNA labeled with genetic markers rather than with radioactive isotopes has the advantage of enabling biologically significant incorporation of exogenous DNA (as determined by expression and heritability) to be distinguished from nonspecific adsorption to cell or organelle membranes. Further, it becomes possible to employ selection methods that permit large numbers of cells to be screened for what is expected to be a rare event. But if genetic markers are to be useful in such selection experiments, they must be chosen judiciously. In microbial transformation systems the incorporation of donor DNA carrying normal alleles of metabolically essential genes is monitored routinely in auxotrophic cells possessing mutant alleles of those genes. But one must have some assurance that transitions to prototrophy result from complementation or correction of the mutant allele of the recipient cell by donor DNA rather than by its reversion. Therefore, it is critical that nonrevertible mutant alleles (e.g., deletions) be employed in such experiments. Unfortunately, none of the presently available mutations of higher plants that are expressed by cultured cells has been shown to satisfy this requirement. However, any novel trait that the donor DNA carries and that the recipient cells lack can be used as a marker to detect incorporation. In such cases it is necessary to establish that appearance of the novel phenotype cannot have been produced by mutation of the recipient cell. Resistances to certain antimetabolites (Table 4.1) are therefore unacceptable as markers in experiments of this type. The ideal genetic markers for use in DNA uptake experiments are single genes that cannot arise by mutation or rearrangement of the recipient cell genome and that are obtained from evolutionarily distant sources. With such markers, incorporation of donor DNA into the recipient cell can be confirmed either by using nucleic acid hybridization studies to demonstrate the presence of the foreign genetic sequence or by identifying the foreign gene product by means of its unique molecular properties.

An attempt to transfer genetic information via the introduction of exogenous DNA into protoplasts was reported by Uchimiya and Murashige (1977). Protoplasts prepared from a *N. tabacum* cultivar that is susceptible to tobacco mosaic virus (TMV) were exposed to DNA isolated from a TMV-resistant cultivar. Of approximately 500 regenerated plants, none displayed resistance to the virus. This result is hardly surprising. The average of 5×10^{-14} g of acid-insoluble DNA taken up by each *N. tabacum* protoplast in these experiments is approximately equivalent to 5×10^{-3} tobacco genomes. Thus, even if all of the incorporated DNA fragments contained intact genes that were replicated, transcribed, and translated (which is highly improbable), a single donor genome would be distributed among a minimum of 200 protoplasts. Admittedly, in screening such a small population the odds are against recovering a protoplast that has incorporated the gene responsible for TMV resistance. Now this experiment, although not yielding the desired result, is not without value, for it serves to illustrate the importance of enriching the population of donor DNA molecules for the sequence of interest and of having a means by which to select those rare cells into which that sequence may have been incorporated.

Recent years have witnessed the development of a powerful new technology for purifying DNA sequences. In this procedure, which is called cloning, restriction endonucleases are used to cleave isolated DNA into fragments. These fragments are then integrated into bacteriophages or plasmids that are capable of replicating autonomously in a bacterial cell. Each bacterial cell contains one fragment of the foreign DNA, which by replicating provides a clone of that sequence. One benefit of this method is that large numbers of the recombinant plasmid (and hence of the desired gene) can be produced simply by multiplication in a bacterial culture. Another advantage of this procedure is that it allows the relative concentration of the desired sequence to be increased dramatically. Cloning a fragment of a haploid plant genome of 5×10^{12} d in a bacterial plasmid of 5×10^6 d effects a millionfold purification of that fragment. Accordingly, by first cloning the desired plant DNA sequence, one reduces by a factor of 10^6 the number of protoplasts that must be examined to recover one into which that sequence has been incorporated.

Polacco, Sparks, and Havir (1979) have described initial efforts to clone the soybean urease structural gene by this method. Fragments produced by endonucleolytic digestion of soybean DNA were integrated into *E. coli* plasmid pBR322, and these recombinant plasmids were then introduced into *E. coli* cells. Since *E. coli* cells normally do not possess a urease gene and therefore cannot utilize urea, it was hoped that bacterial cells containing a plasmid into which the soybean urease gene had been inserted could be selected by their ability to grow when urea was provided as the sole nitrogen source. Although a negative feature of this experimental design was its dependence on normal function of the soybean urease gene and enzyme in the *E. coli* host, if it were successful it would provide the important advantage of identifying the desired recombinant plasmid by direct selection. A few urea-utilizing bacterial clones were recovered in this first venture, but, unfortunately, expression of urea utilization was transitory. Nonetheless, this approach, which in principle could be used for cloning other plant genes, might yet prove feasible.

Genetic markers that under certain circumstances are well suited to studies of DNA uptake are bacterial genes encoding enzymes that modify and thereby inactivate various antibiotics. These genes are available in a highly purified form as components of plasmids called R factors. An *E. coli* plasmid containing an R-factor-derived kanamycin-resistance gene has been employed in an experiment designed to detect transfer of genetic information to *N. tabacum* protoplasts by selecting for function of the bacterial gene in derivative calluses. Protoplasts were incubated in a medium to which had been added either the plasmid bearing the

Table 7.1. *Screening of plantlets derived from* N. tabacum *protoplasts exposed to* E. coli *plasmid DNA for expression of plasmid-encoded kanamycin resistance*

	Treatment with plasmid lacking kanamycin-resistance gene	Treatment with plasmid bearing kanamycin-resistance gene
No. of protoplasts treated	8.2×10^6	7.3×10^6
No. of calluses screened on kanamycin-supplemented medium	326,000	272,000
No. of plantlets regenerated	32	52
Frequency of plantlets regenerated (based on number of calluses screened)	9.8×10^{-5}	19.1×10^{-5}
No. of plantlets tested further for resistance to kanamycin	22	47

Source: Owens (1979).

kanamycin-resistance gene or a related plasmid that lacked this gene (control). When small calluses had developed (13 days after protoplast isolation), the cultures were placed on a shoot-induction medium containing 10 μM kanamycin. Plantlets that did regenerate on this medium were tested further for antibiotic resistance by placing leaf pieces on shoot-inducing medium supplemented with 16 μM kanamycin. After 30 days the dry weight of shoots formed from these explants was determined. A higher frequency of plantlet regeneration was obtained from cultures that had been treated with the plasmid containing the kanamycin-resistance gene than from control cultures (Table 7.1). But as leaf explants isolated from the two plant populations responded similarly to kanamycin, it was concluded that the plasmid-borne gene was not expressed in the tobacco cells (Owens, 1979).

It is quite possible, however, that uptake and expression of the kanamycin-resistance gene do occur, but for any of several reasons could not be observed in these experiments. One consideration is that introduction and expression of a foreign gene occurs at a frequency too low to be detected in the population screened. Mutant *E. coli* cells that lack deoxyribonuclease and therefore are unable to degrade donor DNA are transformed for various single genetic markers at a frequency of only 10^{-6} (Cosloy and Oishi, 1973). Because the number of tobacco calluses examined for expression of kanamycin resistance precludes detection of a frequency below 4×10^{-6}, it is not at all improbable that uptake and expression do occur, but at a frequency below this limit of detection. The procedure by which tobacco calluses were tested for expression of the foreign genetic marker also might be questioned. Resistance to kanamycin was not assayed simply by growth of callus cultures after transfer to a normal callus maintenance medium supplemented with the drug. Rather, an added dimension of complexity and stringency was imposed by demanding expression of the donor marker during differentiation. Since kanamycin interferes with ribosome function, the intense metabolic activity and synthesis of new proteins that are essential to differentiation could make the differentiating plantlet particularly vulnerable to the antibiotic. Hence, a low level of expression of the bacterial kanamycin-resistance gene might be sufficient to effect resistance in the callus, but inadequate to protect the differentiating plantlet. Yet another explanation might be considered for the failure to observe transfer of kanamycin resistance to the tobacco cultures. It is quite conceivable that the acetylated or phosphorylated forms of kanamycin that are produced by bacterial enzymes, although unable to obstruct bacterial ribosomal function, do affect plant mitochondrial or chloroplast ribosomes. It is also possible that plant enzymes are able to restore the modified form of kanamycin to its original and toxic state by essentially reversing the reaction catalyzed by the bacterial enzyme. Thus, if expression of antibiotic resistance is to be regarded as evidence of DNA uptake, it first must be shown that modification of the antibiotic by the bacterial enzyme also eliminates the toxicity of that antibiotic to the plant and that the modified form of the antibiotic is stable within the plant cell. Apropos this point is the demonstration that acetylation of chloramphenicol by the enzyme encoded by an R-factor-derived bacterial resistance gene did not reduce the toxicity of this antibiotic to cultured tobacco cells (R. L. Keil, personal communication). Although these studies did not determine whether acetylated chloramphenicol itself was toxic or was merely deacetylated intracellularly, they did emphasize the necessity of appropriately characterizing the experimental system before drawing conclusions from the results obtained.

Uptake of encapsulated DNA

One of the primary impediments to the introduction of exogenous DNA into plant protoplasts is the presence in the culture medium of nucleases released by the protoplasts (Lurquin and Kado, 1977; Hughes, White, and Smith, 1977, 1978). Some reduction of this nucleolytic activity has been achieved by simply washing protoplasts or by adding proteases (Bendich and Filner, 1971), poly-L-ornithine (Hughes, White, and Smith, 1979), or citrate (Slavik and Widholm, 1978) to the medium and by lowering the incubation temperature (Slavik and Widholm, 1978). Protection against enzymatic digestion may also be afforded by encapsulating exogenously supplied DNA with a membrane or protein coat. DNA encased in liposomes composed of lecithin or of lecithin and cholesterol is largely impervious to the action of deoxyribonuclease. That these liposomes interact readily with protoplasts under conditions that promote protoplast fusion suggests that they may serve as vehicles for the efficient transport of exogenous DNA into protoplasts (Lurquin, 1979).

Protectively packaged DNA is also provided by bacteriophages. Moreover, because of the relatively small size of a bacteriophage genome (the λ genome is 31×10^6 d), bacterial or plant genes are enriched considerably by integration into that genome. The possibility of using bacteriophages as vectors to introduce foreign genetic information into plant cells or protoplasts has been explored in several laboratories. Following exposure of *Hordeum vulgare* protoplasts to the virulent bacteriophage T3, Carlson (1973b) observed synthesis of two phage-specific enzymes. These two enzymes, *S*-adenosylmethionine-cleaving enzyme (SAMase) and RNA polymerase, are encoded by the early region of the T3 genome and normally are not produced by the plant. A bacteriophage origin could definitely be assigned to the latter enzymatic activity, since exposure of protoplasts to a T3 strain containing an amber (nonsense) mutation in the RNA polymerase structural gene resulted in significant synthesis of SAMase activity only. Control experiments demonstrated that the appearance of the two enzyme activities required intact protoplasts and was not due to bacterial contamination.

The inability of tomato *(Lycopersicon esculentum)* callus to use either lactose or galactose as a carbon source provided Doy, Gresshoff, and Rolfe (1973b) with a means of selecting for the introduction and expression of bacterial genes. The capacity to metabolize these sugars was to be introduced by inoculating tomato callus cultures with preparations of specialized transducing phages carrying genes of the *E. coli* galactose *(gal)* and lactose *(lac)* operons. One transducing phage ($\phi80plac^+$) carried the bacterial gene encoding β -galactosidase, the enzyme that catalyzes the hydrolysis of lactose to glucose and galactose. Bacterial genes encoding enzymes that convert galactose to glucose-1-phosphate were carried by another transducing phage *($\lambda pgal^+$)*. In contrast to untreated tomato callus, calluses inoculated with $\lambda pgal^+$ survived and grew slowly on galactose medium. Apparently growth on medium containing galactose as the sole carbon source requires an intact gal^+ operon, since calluses inoculated with $\phi80plac^+$, or $\lambda pgal^-$ were unable to utilize this sugar. Selection for lactose utilization was less stringent. As many as 10 percent of untreated tomato calluses were able to grow on lactose medium.

Inoculation with $\phi 80$ or $\lambda pgal^+$, did not alter this result. However, following inoculation with $\phi 80plac^+$ alone or in combination with $\lambda pgal^+$, 80 percent of the treated callus grew slowly on lactose medium. Immunological identification of the β-galactosidase activity in extracts of the lactose-utilizing callus provided rather convincing evidence of bacterial gene expression. Antiserum specific for the *E. coli* enzyme afforded the same degree of protection against thermal denaturation of the β-galactosidase activity in extracts of both E. *coli* and the lactose-utilizing callus (Table 7.2). Unfortunately, however, expression of the bacterial gene was unstable and ceased on continued propagation of the callus culture (Doy, 1977).

The results of Doy, Gresshoff, and Rolfe (1973b) were exciting, and they stimulated many efforts to repeat them. Although others, by similar means, were able to obtain growth of cultured plant cells on lactose medium, none could demonstrate synthesis of a bacterial enzyme by the plant cells. Inoculation with λ phage carrying the *E. coli lac* genes, but not with wild-type phage (λ^+), enabled sycamore cell suspension cultures to utilize lactose (Johnson, Grierson, and Smith, 1973). Extracts of cultures that had been exposed to $\lambda plac^+$ contained β-galactosidase activity (determined as hydrolysis of the artificial substrate *o*-nitrophenyl-β-D-galactopyranoside). However, lesser but still considerable amounts of this activity were also present in extracts of untreated cells and extracts of cells that had been exposed to λ^+ (Smith et al., 1975). Cultured cells of other plant species also possess β-galactosidase activity (Polacco, Paz, and Carlson, 1974; Hess, Leipoldt, and Illg, 1979). The existence of plant-specific β-galactosidases and the report by Doy, Gresshoff, and Rolfe (1973b) that 5 to 10 percent of control calluses grow on lactose medium indicate that the system is not as tidy as originally supposed. At this point one must ask whether growth on lactose medium is actually accomplished by synthesis of the bacterial enzyme from DNA introduced into the plant cell or merely by induction of enzymes normally encoded by the plant genome. Some evidence suggests that the latter explanation may be correct. A specific inhibitor of *E. coli* β-galactosidase (β-phenyl-thiogalactoside) was shown to have no effect on the activity present in sycamore cells growing in lactose medium following exposure to $\lambda plac^+$. Further, bacterial enzyme could not be detected in these cells by sensitive immunological tests employing antiserum to *E. coli* β-galactosidase (Smith et al., 1975). Using the same highly specific immunological assay for *E. coli* β-galactosidase as Doy, Gresshoff, and Rolfe (1973b), Kapitsa and associates (1979) obtained very different results. Antiserum that protected *E. coli* β-galactosidase against thermal inactivation afforded no protection to the β-galactosidase activity present in tobacco callus or cell suspension cultures that had acquired the

Table 7.2. *Protection by antiserum specific for* E. coli *β-galactosidase of β-galactosidase activity in crude extracts of tomato callus*

Phage inoculation of callus	Carbon source	Growth	β-Galactosidase specific activity (units)	Percentage activity remaining after 20 min at 60°C in presence of rabbit antiserum specific for *E. coli* β-galactosidase
None	Glucose	Rapid	158	<5
None	Lactose	None	144	<5
None	Lactose	Slow (rare)	264	<5
φ80plac⁺ plus λpgal⁺	Lactose	Slow	6850	50–60
None plus *E. coli* extract containing β-galactosidase	Glucose	Rapid	[a]	50–60

[a] An amount of *E. coli* extract was added to give a level of β-galactosidase activity comparable to that in the phage-inoculated callus.
Source: Doy, Gresshoff, and Rolfe (1973b).

Table 7.3. *Thermal inactivation of β-galactosidase activity in crude extracts of tobacco callus in presence of antiserum specific for* E. coli *β-galactosidase*

Source of extract	β-Galactosidase specific activity (relative units)	Percentage activity remaining after 20 min at 61.5°C in presence of rabbit antiserum specific for *E. coli* β-galactosidase
E. coli	0.95	80
E. coli extract containing β-galactosidase plus extract of untreated tobacco callus	1.68	64
Tobacco cell suspension cultures		
Untreated grown on sucrose	0.68	<4
Inoculated with λ*plac*⁺ and grown on lactose	0.88	<4
Tobacco callus cultures		
Untreated grown on sucrose	0.7	<4
Inoculated with λ*plac*⁺ and grown on lactose	0.18	<4
Derived from cell suspension culture inoculated with λ*plac*⁺ and grown on lactose	0.4	<4

Source: Kapitsa et al. (1979).

capacity to utilize lactose after inoculation with λ*plac*⁺ (Table 7.3). Although phage-inoculated tobacco cells growing on lactose contained higher than normal levels of β-galactosidase activity, the pH optimum of this activity was that of the plant (4.4 to 4.7) rather than that of the *E. coli* (7.0 or 5.3) enzyme. These results, in addition to the observation that lactose-utilizing cell lines appear at a low frequency, suggested to Kapitsa, Kulinich, and Vinetskii (1977) that the ability of tobacco cells to grow on lactose medium results from mutation.

These characteristics of the tobacco and sycamore systems clearly distinguish them from the tomato system of Doy, Gresshoff, and Rolfe (1973b). In the latter experiments, most of the calluses inoculated with φ*80plac*⁺ and all of the calluses inoculated with λ*pgal*⁺ acquired the ability to utilize lactose and galactose, respectively. Further, *E. coli* β-galactosidase was identified immunologically in extracts of callus treated with $\phi80plac^+$. All that can be said at this time is that the use of bacteriophages to introduce foreign genes into cultured plant cells remains an exciting (and controversial) possibility. At least we are now aware that in such experiments more dangers of misinterpretation are inherent than were initially apparent.

Other possible vectors

Specialized transducing bacteriophages and bacterial plasmids were first employed as vectors to transfer foreign genetic information into plant cells because at that time they were the sole means by which defined genes could be purified and amplified. However, as these methods of cloning depended on in vivo recombination, they served only to make genetic material from a small number of bacterial species available for gene transfer experiments. Now that enzymatic procedures have been developed for isolating genes from one genome and inserting them into another, DNA from more diverse sources (e.g., plants) can be integrated into phages and plasmids. This technology also frees us to explore the possibility of using other vectors that might be better suited to the task of introducing foreign DNA into plant cells.

To serve as a vector, a DNA molecule must be able to transport intact foreign genetic information into the recipient cell. It must also provide a means for replicating the foreign DNA in the recipient cell. This function can be achieved in either of two ways: the vector can facilitate insertion of the foreign DNA into the genome of the recipient cell, or by possessing an origin of replication (site on the DNA molecule at which replication is initiated), it can enable the recombinant molecule to replicate autonomously in the recipient cell.

Plant viruses may eventually be adapted for use as vectors, but at present they lack certain features that are desirable for such application. Most plant viruses have RNA genomes. Because there is no evidence that any of these RNA viruses is replicated via a DNA intermediate, it is unlikely that they can be maintained stably or can be transmitted predictably in crosses. A small number of plant DNA viruses have been characterized (as reviewed by Kado, 1979), but nothing suggests that any of these DNA viruses are able to integrate into the plant genome. Therefore, such viruses would be useful as vectors only because of their ability to infect and to replicate autonomously in the plant cell. However, as these viruses are produced in only low numbers by the infected plant and hence are difficult to purify, it might be preferable to integrate a viral origin of replication into a bacterial plasmid and thereby confer on the plasmid the capacity to replicate within a plant cell. This approach could offer additional advantages. Such a hybrid bacterial plasmid might not have the limited host range nor produce in the regenerated plant the pathogenic symptoms that are characteristic of the plant virus.

As DNA molecules that are much smaller than the nuclear genome and that can replicate within a plant cell, mitochondrial and chloroplast chromosomes may also be made to serve as vectors for transferring foreign genetic information. Isolated organelle DNA into which additional sequences have been inserted enzymatically in vitro might be taken up directly by protoplasts. However, as protoplasts are able to incorporate intact chloroplasts (see Chapter 6), a modified organelle genome might first be repackaged within a membrane before being mixed with a population of protoplasts. The peculiar features of an organelle genome that make it attractive as a possible vector could also impose limitations on its usefulness. We do not know if nuclear genes inserted into the organelle chromosome would be transcribed or translated in the organelle and, if they were, whether or not the polypeptide products would ever be transported to the cytoplasm.

Plants contain transposable elements that might also be exploited for introducing foreign DNA into cultured cells. Although these elements are not yet well characterized, they are believed to be DNA sequences capable of promoting their own insertion into and excision from plant chromosomes (as reviewed by Fincham and Sastry, 1974). If plant transposable elements are similar to bacterial transposons, then the ability of transposable elements to insert into plant chromosomes could facilitate the replication, transcription, and inheritance of foreign sequences that have been attached to them. On the other hand, the facility of these elements for spontaneous excision could prove a serious disadvantage.

Crown gall is a disease of plants that has long been known to be caused by *Agrobacterium tumefaciens*. Interest in crown gall was first aroused by the formation of tumors by which this disease is characterized. However, hopes that virulence (Ti) plasmids of *A. tumefaciens* could be used to transport foreign genes into plant cells came with the discoveries that tumorigenesis results from the transfer into plant cells of part of a Ti plasmid (Chilton et al., 1977) and that this bacterial DNA is transcribed in tumorous cells (Drummond et al., 1977). An important step toward this goal has recently been made with the successful in vitro transformation of cultured tobacco cells by virulent strains of *A. tumefaciens*. Shoots regenerated from several transformed callus cultures have been mor-phologically abnormal and have contained lysopine dehydrogenase, an enzyme that is found only in transformed tissues (Márton et al., 1979). But three critical features of the crown gall system may obstruct or limit the use of Ti plasmids for introducing foreign DNA into cultured plant cells. First, the host range of *A. tumefaciens* is restricted to dicotyledonous species. Second, it will be necessary to separate the regions of the Ti plasmid that promote its transfer into the plant cell from those that cause pathogenic effects, such as tumorigenesis, if indeed these functions are encoded by distinct segments of the plasmid. However, some basis for optimism is provided by the finding that not all of the transformed calluses produced in vitro by Márton and associates (1979) expressed both of the characteristic biochemical markers of crown gall tumors that were examined. Some transformed calluses were capable of growth in the absence of exogenous hormones (a property associated with tumor cells), but did not possess lysopine dehydrogenase, whereas another callus that did produce lysopine dehydrogenase required exogenous hormones for growth. The third difficulty may present the greatest challenge. For the Ti plasmid to serve as a vector for introducing foreign DNA, it must be maintained stably in the plant cells. But neither progeny of plants regenerated from transformed tissues nor haploid individuals produced by culturing anthers of such plants have manifested neoplastic or biochemical properties of crown gall tumor tissue (Turgeon, Wood, and Braun, 1976). Apparently, during meiosis either the Ti plasmid DNA (or at least those sequences causing tumor formation) is eliminated from the plant cells or its expression is completely suppressed. Of course, if suppression of plasmid DNA expression occurs, a means might be found for its reversal. And if plasmid DNA as such is eliminated during meiosis, other sequences that it carries may not necessarily be eliminated as well.

Although the introduction of purified DNA into cultured plant cells offers exciting prospects for the genetic engineering of plants, the exploration of this approach is only just beginning. The brevity of this chapter reflects the small amount of relevant research that has been performed to date, rather than the promise that this approach holds. I am confident that in any future edition of this book the proportionate increase in length of this chapter will be greater than that of any other.

8

Applications to plant breeding

Most fields of scientific inquiry arise from the necessity of discovering solutions to practical problems. Thus the desire to harness the power of steam gave rise to the science of thermodynamics, and the urgency of easing human suffering has always provided the impetus for the development of medical science. Similarly, man's constant striving to improve the quality and yield of plants on which he depends for his subsistence first established the science of plant breeding and now that of plant cell genetics.

Yet knowledge acquired in pursuit of practical goals may open new lines of questioning. Such tangents often develop as areas of investigation that are intellectually valid in their own right, although all too frequently the memory of their origin fades during the course of this development. One can speculate that the sciences of plant physiology, biophysics, and molecular genetics arose in this way. It is essential that these more "basic" sciences be encouraged and supported even though no practical benefit from the knowledge acquired is immediately obvious. Often it is impossible to predict when and in what form such research will yield results that represent a significant contribution to the "applied" science from which it sprang, but it is inevitable that it will do so. Molecular genetics was criticized in the early 1970s because of its failure up to that time to make any tangible contribution to human welfare. But now amniocentesis has become a routine procedure for accomplishing prenatal diagnosis of genetic diseases, and the technology of recombinant DNA has revolutionized the medically important manufacture of peptide hormones and promises exciting applications to industrial fermentation and plant breeding (see Chapter 7). Clearly the dichotomy that has been established between basic science and applied science is a false and dangerous one. It is as myopic and self-defeating for those in the applied sciences to attempt to stifle development of basic research as it is for those in basic research to scorn the accomplishments of applied research. Here I will enter a plea that basic science and applied science no longer be considered as separate and antagonistic endeavors. Progress will be achieved only through the realization that both sciences interact and

148

develop together dialectically and that each flows from and contributes to the development and shaping of the other.

Perhaps the greatest danger faced by the fledgling field of plant cell genetics is the premature and unrealistic imposition of demands for its application to problems of plant breeding. Although practical application is undoubtedly the ultimate ambition of all those working in the field, untimely demands may stunt the development of the infant science and prevent the establishment of a mature and independent science that eventually would be able to satisfy such demands as well as make contributions not now foreseen. Certainly a few applications may be anticipated in the near future. But many basic problems, such as controllable regeneration of plants from single protoplasts of crop species, must be solved before the full potentiality of plant cell genetics is realized. Thus, although a primary motivation to plant cell genetics research is its practical application, an enormous amount of preliminary work is needed to develop these methods to the point where such application becomes possible.

Plant breeding is a pragmatic and largely empirical endeavor. In a conventional breeding program a hybrid is first constructed by crossing genetically different plants. The hybrid is then crossed, and desirable recombinant genotypes are identified among the segregating progeny. When the sources of variability have been exhausted and all of the genetic permutations have been tried, it is announced that a "plateau" has been reached. But of course such a cul-de-sac is expected when further manipulation of these traits is precluded by lack of knowledge of and inability to modify directly the underlying genetic and biochemical processes. It is by providing this knowledge and the means by which to alter these processes in a defined way that cell genetics will contribute to plant breeding.

Limitations of an in vitro approach

Specific potential applications of each of the several techniques of cell and protoplast culture have been enumerated in the appropriate chapters. Advances realized in plant cell culture in the past few years extend considerably beyond the mere selection of variant phenotypes. Cells can be separated into their component organelles and macromolecules. Other techniques have been developed for the modification and transfer of these components. Finally, novel cell types can be constructed by bringing together these subcellular components in new combinations. But, as discussed in Chapter 2, plant cell genetics by its very nature is severely limited in its ability to contribute directly to the solution of agronomic problems. Many of the limitations of a cell culture system can be avoided if one is willing to forego the use of selection methods. Providing that plants can be regenerated from protoplasts, there is no limitation, save that of human endurance, that prevents one from regenerating large populations of plants following mutagenesis or protoplast fusion and screening for expression of the desired phenotype by conventional means. Most of the limitations of cell culture for genetic modification of plants are encountered when one employs selection schemes that require expression of particular functions by cultured cells. It is as if a Faustian bargain has determined that certain experimental constraints be accepted in return for the convenience of selection procedures.

The first and perhaps most obvious limitation of an in vitro approach is that it can be used to select only for alterations of traits that are expressed by cultured cells. Thus, modifications of many important whole-plant features such as root and leaf morphology, yield, and maturation time cannot be selected directly in vitro. Such traits are properties of differentiated organ systems not present in cell cultures, and the functions corresponding to these traits that might be expressed by cultured cells have not yet been identified. But that is not to say that the expression of differentiated functions by cultured cells cannot be accomplished. Recent definition of conditions for photoautotrophic growth of cultured cells (Berlyn and Zelitch, 1975; Berlyn, Zelitch, and Beaudette, 1978; Yamada, Sato, and Hagimori, 1978) has made possible the application of in vitro techniques to study and perhaps genetically manipulate the processes of photosynthesis and photorespiration.

If the function in which variation is being sought is expressed by cultured plant cells, one must next devise methods for selecting those cells possessing the desired modification of that function. This task is trivial when the desired phenotype is increased resistance to a toxic compound that can be incorporated into the medium (e.g., heavy metals or herbicides). But other cases pose a challenge to one's ingenuity. For example, it has been demonstrated that under precisely defined conditions *Phytophthora parasitica* var. *nicotianae,* the causative agent of black shank disease, will grow on callus cultures derived from susceptible, but not on those derived from resistant, varieties of *Nicotiana tabacum* (Helgeson et al., 1972). Yet it seems doubtful that infection of susceptible callus cultures with the fungus can be used to select for resistance to black shank disease. Because the response of the callus culture to the fungus is expressed at the level of the cell population rather than at the level of the individual cell, a small number of resistant cells within a predominantly susceptible population will go undetected. Insect resistance is another example of a trait for which selection in vitro will prove difficult. Cultured cells might synthesize compounds that would act as insect repellents in the whole plant, but obviously one cannot introduce insects into a Petri dish to select those variant cells that produce elevated levels of these compounds. On the other hand, elucidation of the sequence and mechanism of regulation of biosynthesis of the repellent might lead to the development of a means for selecting directly for its increased rate of production. When this biochemical information is available, a medium that will provide a selective growth advantage to cells of the desired genotype can be formulated. Indeed, it is ignorance of the genetic, biochemical, and physiological processes responsible for plant characteristics that most often frustrates efforts to design procedures for selecting cells in which that characteristic has been modified or introduced by gene transfer. What is the molecular basis of nematode resistance? Of cold and drought tolerance? And how can modifications of these traits be selected at the cellular level if the appropriate gene products have not been identified at that level?

A third consideration for those who would use cell culture to modify whole-plant characteristics is whether or not functions expressed in vitro will be expressed in the mature plant. Enormous effort could be expended in selection of a particular phenotype in culture only to discover that the alteration is not manifested in the whole plant, or at least not in the manner hoped for (see Chapter 2). A case that illustrates this point as well as some of the potential advantages of cell culture is the endeavor to select in vitro mutations that modify the protein quality of the seed.

The conventional method of isolating mutants or varieties that contain increased quantities of protein or protein of improved nutritional quality is extremely laborious. By this method, plants must be grown to maturity and the seeds of each genotype harvested and assayed. In the case of cereals, the magnitude of this task was lessened considerably by the discovery of two mutants of maize *(opaque-2* and *floury-2),* which established an association between improved lysine content and the floury endosperm phenotype (Mertz, Bates, and Nelson, 1964; Nelson, Mertz, and Bates, 1965). High-lysine varieties can now be identified more efficiently by choosing for amino acid analysis only those varieties of a grain that possess a particular endosperm morphology (Singh and Axtell, 1973). But a great price must be paid for this convenience. Recovering only those high-lysine mutants that possess a floury endosperm narrows considerably the spectrum of desirable mutant types that can be isolated. Further, the floury endosperm is an undesirable trait that may be difficult or impossible to separate genetically from increased lysine content.

In contrast, the in vitro approach to the improvement of protein quality selects directly for a modification of the control of amino acid biosynthesis. In this procedure cultured cells are incubated in the presence of amino acid analogues or growth-inhibitory mixtures of amino acids, and resistant cell lines are isolated (see Chapter 4). The question, of course, is whether or not an increased rate of synthesis of a particular amino acid in the cultured cell will lead to an increased quantity of that amino acid in the seed protein in situ. For several years speculation in both directions was rife, but the answer was obtainable only by experiment. Now it has been reported that selection among cultured cells for resistance to a growth-inhibitory mixture of lysine plus threonine has resulted in the isolation of a maize mutant that contains an increased amount of threonine in the kernel (K. A. Hibberd and C. E. Green, personal communication). Doubtless, as more is learned about the regulation of amino acid biosynthesis in plants (in part through the isolation of such mutants in vitro), procedures will be devised for selecting additional mutations affecting the production and accumulation of other amino acids. Yet by insisting a priori that the amino acid composition of the seed would not be affected by mutations selected in vitro, some scientists would have closed this avenue to the improvement of the nutritional value of crop plants.

Thus far we have considered an approach to the improvement of seed protein quality that is based on selection of mutants that produce increased quantities of specific amino acids. As amino acids are the substrates for the synthesis of proteins, it is not unreasonable to suppose that, providing the supply of a particular amino acid is limiting to the synthesis of proteins in which it is contained, increased amounts of such proteins will be produced in response to an increased supply of that amino acid. An alternative and perhaps more direct method of altering the amino acid composition of the seed is possible: one could devise schemes for selecting in vitro regulatory mutations that derepress synthesis of specific seed proteins containing above-average amounts of nutritionally limiting amino acids. However, since mutant selection strategies most often are based on establishing a growth differential between normal and mutant cells, only regulatory mutants affecting seed proteins possessing a catalytic activity could be selected. This activity would be exploited by the selection procedure to achieve the requisite growth differential. Procedures for identifying mutations affecting the rate of synthesis of nonenzymatic proteins, such as storage proteins, would be much more difficult to devise.

Urease was recognized by Polacco (1976) as an enzyme present in legume seeds that is potentially well suited to such genetic manipulation. Relative to other proteins in soybean seeds, urease contains a large amount of methionine, which is the amino acid limiting to the nutritional quality of soybean protein (Polacco, Sparks, and Havir, 1979). Polacco (1976) studied the regulation of urease activity in cultured soybean cells in order to define conditions for selecting mutants that synthesize more than the normal amounts of this enzyme. These studies revealed that utilization of urea as a nitrogen source by cultured soybean cells can be prevented by hydroxyurea, by nitrate, or by methylammonia. Selection for cells able to utilize urea in the presence of any one of these compounds might result in the isolation of mutants in which the regulation of urease synthesis is altered. But even this approach really offers little more certainty than that of selecting for increased production of amino acids that the mutant trait will be expressed in the seed. Although urease normally is produced in the seed, we do not know that its synthesis is regulated by the same mechanism in that organ as in cell culture. Hence, mutations affecting the control of urease synthesis in vitro may not alter its rate of synthesis in vivo. We are left where we began: unable to predict whether or not a novel phenotype selected in cell culture will be expressed in the mature plant. The experiment must be performed in each case. There is no getting around it. Indeed, speculation about the feasibility of an in vitro approach could very well prove harmful and self-defeating. Predictions of its impracticability could delay the execution of what is in fact a valuable line of experimentation.

In vitro manipulations not involving selection at the cellular level

The primary advantages of plant cell and tissue culture for crop improvement that have been emphasized in this book are that it makes possible the generation of enormous genetic variability in a small volume, the introduction of procedures for selecting desirable variants, and the performance of genetic manipulations that cannot be achieved by conventional means. However, other applications of tissue culture also merit the attention of the plant breeder. The ability to regenerate plants at will from tissues propagated in vitro permits rapid multiplication of a desirable genotype. This technique of vegetative propagation is employed routinely by the horticultural industry and is of increasing importance in tree production. The tendency of meristematic tissues to exclude pathogens provides the basis for another commercial application of plant tissue culture. Thus the differentiation of cultured meristematic tissues is employed widely to produce disease-free plants, but because these techniques are not useful for genetic manipulation, they have not been considered in this volume. However, genetic manipulation of plants can be achieved by pollen and anther culture, and therefore some discussion of these techniques is appropriate at this point. The details of these techniques, by which plants can be induced to regenerate from the male gametes by culturing isolated immature pollen or anthers of some species, are described in a recent review by Sunderland and Dunwell (1977). The genetic consequence of anther and pollen culture of interest here is that homozygous diploid plants are produced directly from the haploid meiotic segregants by either spontaneous endomitosis (or endoreduplication) or treatment with colchicine. In contrast, the attainment of homozygosity in conventional pedigree and backcross breeding programs might require five to six generations. But continued selection and recombination of genetic material in these successive generations are advantages of the conventional method that are not offered by anther culture. For example, in a backcross breeding program designed to incorporate a single desirable gene into an established cultivar, repeated crosses are made to the recurrent parent, and then this last generation is selfed to produce completely homozygous individuals. If, on the other hand, anthers of the F_1 plant are cultured, homozygous diploids can be derived directly from the gametes of this hybrid, and one need only identify an individual that obtained its entire chromosome complement from the established cultivar except for the one chromosome carrying the single gene whose introduction is desired. But if deleterious genes are linked to the desirable one, it becomes necessary to perform additional crosses and screen for progeny from which the deleterious genes have been eliminated by recombination. Thus it might be preferable to culture anthers of the second- or third-generation plants obtained from a conventional breeding program. But at this point the primary advantage of anther culture, which lies in reducing the time required to produce the desired homozygote, becomes less apparent.

Another factor to be considered in employing anther culture for breeding is the dependence of the success of this technique on the genotype of the parent plant. It has been shown, for example, that a high proportion of anthers of some varieties of rice produce callus or plantlets in response to being placed in culture, whereas anthers of other varieties do so rarely or not at all (Guha-Mukherjee, 1973; Oono, 1975; Chen and Lin, 1976). Further, the proportion of regenerated plants that are albino is also determined primarily by the genotype (Oono, 1975). Similarly, an influence of plant genotype on the frequency of androgenesis has been demonstrated for cultured anthers of rye (Wenzel, Hoffmann, and Thomas, 1977) and potato (Sopory and Rogan, 1977). Hence, in developing a breeding program that makes use of anther culture, one must consider not only the agronomic qualities of the parental varieties but also their responsiveness to tissue culture. Because responsiveness to tissue culture was hardly a property that was deemed important in constructing current crop cultivars, such a requirement will probably prove severely limiting in practice. Therefore, for anther culture to be used as a breeding technique for many crop species, it first will be necessary to screen germ plasm collections for both the frequency of callus induction (or embryogenesis) in cultured anthers and the frequency of regeneration of green plants. These capabilities then will have to be crossed into important cultivars. Perhaps in future breeding programs responsiveness to anther culture will be as central a consideration as yield.

Several cases have been reported of successful application of anther culture to the production of new crop varieties. In the People's Republic of China, varieties of rice, wheat, and tobacco have been developed by use of anther culture (Han et al., 1978). And in the United States (Collins and Legg, 1974) and Japan (Nakamura et al., 1974), anther culture has been used extensively in tobacco breeding programs. Homozygous tobacco lines produced by doubling the chromosome number of haploids derived from anther culture have been found to differ from the parental inbred line in several traits that have been examined. Many haploid-derived tobacco lines are also less vigorous than the parental line. But in no case has significant variation been found within a line produced by anther culture (Devreux and Laneri, 1974; Collins and Legg, 1974; Burk and Matzinger, 1976). Variation between homozygous lines produced by anther culture could be due to segregation of residual heterozygosity of the inbred line from which the haploids are derived. As genetic changes are known to occur spontaneously in cultured cells (see Chapter 3), such variability could also arise during the period of anther culture. Segregation and mutation as well as inbreeding depression could also contribute to the reduced vigor of haploid-derived lines.

A potential application of anther culture that has not yet been explored is in providing material for the construction of superior hybrids. Anther culture offers the possibility of rapidly generating large numbers of genetically distinct homozygous lines for evaluation in hybrid production. Haploids of some plant species obtained by other means (Chase, 1974; Kasha, 1974a,b; Lacadena, 1974) have proved useful for this purpose. The application of anther culture will enlarge the number of species for which the rapid production of homozygous lines is possible.

Certainly anther culture is promising as a technique for use in plant breeding, but this discussion makes evident certain limitations to this utility. Not the least of these limitations is that to date haploid plant production has been accomplished from anthers of but a small number of crop species. Most efforts to define conditions for promoting the development of haploid plants from anthers of more species have been through modification of the medium composition or treatment of plants and detached flowers prior to excision of the anthers. Although the successes to date have been realized primarily through this approach, it now appears to have been largely exhausted. At this time the identification of the genetic factors conditioning responsiveness to anther culture and the crossing of these factors into cultivated varieties promise to be more efficacious procedures.

Inability to control the regeneration of plants from cultured cells is a problem not only with cultured pollen but also with most types of cultured cells. The loss of morphogenetic capacity that occurs on propagation of cell cultures is the single most intractable and frustrating obstacle to the application of in vitro methods to plant breeding. And here, too, the genetic approach appears to be the most promising. In a study of 16 varieties of *Pisum sativum,* Malmberg (1979b) found that callus cultures derived from these varieties differed both in their capacity to give rise to plants and in the period in culture during which this capacity was retained.

While working toward improvement of the tissue culture system for a given crop plant, many useful experiments still can be performed without having developed the ideal system, which would permit plants to be regenerated from single haploid protoplasts. For example, providing that an adequate means of selection is available, one should be able to isolate mutants from callus masses without enjoying the benefit of a single-cell suspension. And in cases in which the morphogenetic capacity of the cell culture continuously diminishes, it may be possible to regenerate plants before that capacity is lost if selection is accomplished quickly. This reasoning has sustained my own hopes of selecting mutants of rice by use of an anther culture system. Although the capacity of rice callus cultures to regenerate plants is lost within three passages in culture, the anther culture system offers large numbers of initially totipotent haploid cells from which mutants can be selected (Chaleff, 1980b).

There can be no doubt that plant cell and tissue culture will afford an important technology for crop improvement, but we must be willing to postpone the rewards and invest time and effort in the development of this technology. Premature expectations could prove destructive: by attempting to reap the crop before it ripens, we risk forfeiting the harvest.

Bibliography

Aarnes, H. (1977). A lysine-sensitive aspartate kinase and two molecular forms of homoserine dehydrogenase from barley seedlings. *Plant Sci. Lett. 9:*137–45.

Alfermann, W., and Reinhard, E. (1971). Isolierung anthocyanhaltiger und anthocyanfreier Gewebestämme von *Daucus carota:* Einfluss von Auxinen auf die Anthocyanbildung. *Experientia 27:*353–4.

Arya, H. C., Hildebrandt, A. C., and Riker, A. J. (1962). Clonal variation of grape-stem and *Phylloxera*-gall callus growing *in vitro* in different concentrations of sugars. *Am. J. Bot. 49:*368–72.

Asada, K., Saito, K., Kitoh, S., and Kasai, Z. (1965). Photosynthesis of glycine and serine in green plants. *Plant Cell Physiol. 6:*47–59.

Aspinall, G. O., Molloy, J. A., and Craig, J. W. T. (1969). Extracellular polysaccharides from suspension-cultured sycamore cells. *Can. J. Biochem. 47:*1063–70.

Aviv, D., Fluhr, R., Edelman, M., and Galun, E. (1980). Progeny analysis of the interspecific somatic hybrids: *Nicotiana tabacum* (CMS) + *Nicotiana sylvestris* with respect to nuclear and chloroplast markers. *Theor. Appl. Genet. 56:*145–50.

Aviv, D., and Galun, E. (1977). Isolation of tobacco protoplasts in the presence of isopropyl-*N*-phenylcarbamate. *Z. Pflanzenphysiol. 83:*267–73.

Bantle, J. A., and Hahn, W. E. (1976). Complexity and characterization of polyadenylated RNA in the mouse brain. *Cell 8:*139–50.

Barg, R., and Umiel, N. (1977). Development of tobacco seedlings and callus cultures in the presence of amitrole. *Z. Pflanzenphysiol. 83:*437–47.

Bateson, W., and Saunders, E. R. (1902). Experimental studies in the physiology of heredity. Reports to the Evolution Committee of the Royal Society. Report 1.

Bateson, W., Saunders, E. R., Punnett, R. C., and Hurst, C. C. (1905). Experimental studies in the physiology of heredity. Reports to the Evolution Committee of the Royal Society. Report 2.

Bayliss, M. W. (1973). Origin of chromosome number variation in cultured plant cells. *Nature (London) 246:*529–30.

Bayliss, M. W. (1975). The effects of growth *in vitro* on the chromosome complement of *Daucus carota* (L.) suspension cultures. *Chromosoma 51:*401–11.

Becker, G. E., Hui, P. A., and Albersheim, P. (1964). Synthesis of extracellular polysaccharide by suspensions of *Acer pseudoplatanus* cells. *Plant Physiol. 39:*913–20.

Behnke, M. (1979). Selection of potato callus for resistance to culture filtrates of *Phytophthora infestans* and regeneration of resistant plants. *Theor. Appl. Genet. 55:*69–71.

Behrend, J., and Mateles, R. (1975). Nitrogen metabolism in plant cell suspension cultures. 1. Effect of amino acids on growth. *Plant Physiol. 56:*584–9.

157

Beier, H., Bruening, G., Russell, M. L., and Tucker, C. L. (1979). Replication of cowpea mosaic virus in protoplasts isolated from immune lines of cowpeas. *Virology 95:*165–75.

Beier, H., Siler, D. J., Russell, M. L., and Bruening, G. (1977). Survey of susceptibility to cowpea mosaic virus among protoplasts and intact plants from *Vigna sinensis* lines. *Phytopathology 67:*917–21.

Belliard, G., Pelletier, G., and Ferault, M. (1977). Fusion de protoplastes de *Nicotiana tabacum* à cytoplasmes différents: etude des hybrides cytoplasmiques néo-formés. *C. R. Acad. Sci. Ser. D 284:*749–52.

Belliard, G., Vedel, F., and Pelletier, G. (1979). Mitochondrial recombination in cytoplasmic hybrids of *Nicotiana tabacum* by protoplast fusion. *Nature (London) 281:*401–3.

Bendich, A. J., and Filner, P. (1971). Uptake of exogenous DNA by pea seedlings and tobacco cells. *Mutat. Res. 13:*199–214.

Berlin, J., and Widholm, J. M. (1977). Correlation between phenylalanine ammonia lyase activity and phenolic biosynthesis in p-fluorophenylalanine-sensitive and -resistant tobacco and carrot tissue cultures. *Plant Physiol. 59:*550–3.

Berlyn, M. B., and Zelitch, I. (1975). Photoautotrophic growth and photosynthesis in tobacco callus cells. *Plant Physiol. 56:*752–6.

Berlyn, M. B., Zelitch, I., and Beaudette, P. D. (1978). Photosynthetic characteristics of photoautotrophically grown tobacco callus cells. *Plant Physiol. 61:*606–10.

Bick, M. D., and Davidson, R. I. (1974). Total substitution of bromodeoxyuridine for thymidine in the DNA of a bromodeoxyuridine-dependent cell line. *Proc. Natl. Acad. Sci. U.S.A. 71:*2082–6.

Binding, H. (1972). Selektion in Kalluskulturen mit haploiden Zellen. *Z. Pflanzenzücht. 67:*33–8.

Binding, H., Binding, K., and Straub, J. (1970). Selektion in Gewebekulturen mit haploiden Zellen. *Naturwissenschaften 3:*138–9.

Binding, H., and Kollmann, R. (1976). The use of subprotoplasts for organelle transplantation. In *Cell Genetics in Higher Plants,* ed. D. Dudits, G. I. Farkas, and P. Maliga, pp. 191–206. Akadémiai Kiadó, Budapest.

Binns, A., and Meins, F. (1973). Habituation of tobacco pith cells for factors promoting cell division is heritable and potentially reversible. *Proc. Natl. Acad. Sci. U.S.A. 70:*2660–2.

Bird, I. F., Cornelius, M. J., Keys, A. J., and Whittingham, C. P. (1972). Oxidation and phosphorylation associated with the conversion of glycine to serine. *Phytochemistry 11:*1587–94.

Blakely, I. M., and Steward, F. C. (1961). Growth induction in cultures of *Haplopappus gracilis.* 1. The behavior of the cultured cells. *Am. J. Bot. 48:*351–8.

Bonnett, H. T., and Eriksson, T. (1974). Transfer of algal chloroplasts into protoplasts of higher plants. *Planta 120:*71–9.

Bourgin, J. P. (1976). Valine-induced inhibition of growth of haploid tobacco protoplasts and its reversal by isoleucine. *Z. Naturforsch. 31c:*337–8.

Bourgin, J. P. (1978). Valine-resistant plants from *in vitro* selected tobacco cells. *Mol. Gen. Genet. 161:*225–30.

Braun, A. C. (1937). A comparative study of *Bacterium tabacum* Wolf and Foster and *Bacterium angulatum* Fromme and Murray. *Phytopathology 27:*283–304.

Braun, A. C. (1955). A study on the mode of action of the wildfire toxin. *Phytopathology 45:*659–64.

oplast regeneration from
allus. *J. Cell Sci. 16*:445–

ficiency of hypoxanthine
ant to azaguanine. *Planta*

Aspartate kinase and the
Planta 139:119–24.
The effect of aspartate-
on the growth of excised

istry of cancer (metabolic

its regulation. In *Plant*
25–60. Academic Press,

n, J. K. (1970). Isolation
ase from a multicellular

g anther-derived doubled
1–4.
toplast-derived haploid
m. *Planta 149*:7–18.
two forms of anthranilate
–resistant cultured *Sol-*

ts in ferns. *Genet. Res.*

phic mutants in somatic

ant mutants of tobacco.

ic research. *Proc. Natl.*

her plant. *Genet. Res.*

Parasexual interspecific
92–4.
N. H. (1979). Efficient
id plasmid *DNA. Proc.*

am-tolerant mutants of

In press.
netics of higher plants.

n for mutants of higher
l, ed. L. Ledoux, pp.

lycerol-utilizing mutant

n *in vitro* for herbicide-

resistant mutants of *Nicotiana t*
75:5104–7.

Chaleff, R. S., and Polacco, J. C. (19
In *The Molecular Biology of Plant*
Scientific Publications, Oxford.

Chase, S. S. (1974). Utilization of h
species. In *Haploids in Higher Pla*
pp. 211–30. University of Guelph,

Chattoo, B. B., Sherman, F., Azuba
Ogur, M. (1979). Selection of *ly*
cerevisiae by the utilization of α-a

Chen, C. C., and Lin, M. H. (197
culture. *Bot. Bull. Acad. Sin.* 17:18

Chen, K., Wildman, S. G., and Smi
tion in parasexual hybrids as sho
protein. *Proc. Natl. Acad. Sci. U*

Cheshire, R. M., and Miflin, B. J.
maize. *Phytochemistry* 14:695–8.

Chilton, M. D., Drummond, M. H.
Gordon, M. P., and Nester, E.
DNA into higher plant cells:
genesis. *Cell* 11:263–71.

Chu, E. H. Y., Brimer, P., Jaco
Mammalian cell genetics 1. Selec
trophic for L-glutamine or resi
cells *in vitro*. *Genetics* 62:359–77.

Chu, E. H. Y., and Powell, S. S
genetics. In *Advances in Human*
vol. 7, pp. 189–258. Plenum Press

Clausen, R. E., and Cameron, D. R.
Monosomic analysis. *Genetics* 29:

Cocking, E. C. (1960). A method
vacuoles. *Nature (London)* 187:9

Cocking, E. C., George, D., Price-J
procedures for the production
hybrida and *Petunia parodii*. 2.
Lett. 10:7–12.

Coen, D. M., Bedbrook, J. R.,
chloroplast DNA fragment encod
carboxylase. *Proc. Natl. Acad. Sc*

Collins, G. B., and Legg, P. D
allopolyploid species. In *Haploi*
ed. K. Kasha, pp. 231–47. Unive

Collins, G. B., Vian, W. E., and
trichloropicolinic acid as an au
18:286–8.

Conde, M. F., Boynton, J. E., Gill
Wang, W. L. (1975). Chloroplast
ribosomes. *Mol. Gen. Genet.* 140

Constabel, F. (1967). Pigmentbild
wissenschaften 54:175–6.

Constabel, F. (1976). Somatic hybri

Constabel, F. (1978). Development of protoplast fusion products, heterokaryo-cytes, and hybrid cells. In *Frontiers of Plant Tissue Culture 1978*, ed. T. A. Thorpe, pp. 141–9. International Association for Plant Tissue Culture, Calgary.

Cosloy, S. D., and Oishi, M. (1973). The nature of the transformation process in *Escherichia coli* K12. *Mol. Gen. Genet. 124:*1–10.

Cove, D. J. (1976). Chlorate toxicity in *Aspergillus nidulans*. *Mol. Gen. Genet. 146:*147–59.

Croughan, T. P., Stavarek, S. J., and Rains, D. W. (1978). Selection of a NaCl-tolerant line of cultured alfalfa cells. *Crop Sci. 18:*959–63.

D'Amato, F. (1977). Cytogenetics of differentiation in tissue and cell cultures. In *Plant Cell, Tissue, and Organ Culture*, ed. J. Reinert and Y. P. S. Bajaj, pp. 343–57. Springer-Verlag, Berlin.

D'Amato, F. (1978). Chromosome number variation in cultured cells and regener-ated plants. In *Frontiers of Plant Tissue Culture 1978*, ed. T. A. Thorpe, pp. 287–95. International Association for Plant Tissue Culture, Calgary.

Davey, M. R., Fowler, M. W., and Street, H. E. (1971). Cell clones contrasted in growth, morphology, and pigmentation isolated from a callus culture of *Atropa belladonna* var. *lutea*. *Phytochemistry 10:*2559–75.

Davey, M. R., Frearson, E. M., and Power, J. B. (1976). Polyethylene glycol-induced transplantation of chloroplasts into protoplasts: an ultrastructural assessment. *Plant Sci. Lett. 7:*7–16.

Davidson, R. L., and Bick, M. D. (1973). Bromodeoxyuridine dependence—a new mutation in mammalian cells. *Proc. Natl. Acad. Sci. U.S.A. 70:*138–42.

DeMars, R., and Hooper, J. L. (1960). A method of selecting for auxotrophic mutants of HeLa cells. *J. Exp. Med. 111:*559–72.

DeMarsac, N. T., and Jouanneau, J. P. (1972). Variation de l'exigence en cytokinine de lignées clonales de cellules de tabac. *Physiol. Veg. 10:*369–80.

Devreux, M., and Laneri, U. (1974). Anther culture, haploid plant, isogenic line, and breeding researches in *Nicotiana tabacum* L. In *Polyploidy and Induced Mutations in Plant Breeding*, pp. 101–7. International Atomic Energy Agency, Vienna.

DeVries, H. (1910). *The Mutation Theory, vol. 2. The Origin of Varieties by Mutation*. Open Court Publishing, Chicago.

DiCamelli, C. A., and Bryan, J. K. (1975). Changes in enzyme regulation during growth of maize. 2. Relationships among multiple molecular forms of homoserine dehydrogenase. *Plant Physiol. 55:*999–1005.

Dix, P. J. (1977). Chilling resistance is not transmitted sexually in plants regenerated from *Nicotiana sylvestris* cell lines. *Z. Pflanzenphysiol. 84:*223–6.

Dix, P. J., Joó, F., and Maliga, P. (1977). A cell line of *Nicotiana sylvestris* with resistance to kanamycin and streptomycin. *Mol. Gen. Genet. 157:*285–90.

Dix, P. J., and Street, H. E. (1974). Effects of *p*-fluorophenylalanine (PFP) on the growth of cell lines differing in ploidy and derived from *Nicotiana syl-vestris*. *Plant Sci. Lett. 3:*283–8.

Dix, P. J., and Street, H. E. (1975). Sodium chloride-resistant cultured cell lines from *Nicotiana sylvestris* and *Capsicum annuum*. *Plant Sci. Lett. 5:*231–7.

Dix, P. J., and Street, H. E. (1976). Selection of plant cell lines with enhanced chilling resistance. *Ann. Bot. (London) 40:*903–10.

Doy, C. H. (1977). Phage-mediated transgenosis in plant cells. In *Molecular Genetic Modification of Eucaryotes*, ed. I. Rubenstein, R. L. Phillips, C. E. Green, and R. J. Desnick, pp. 133–5. Academic Press, New York.

Doy, C. H., Gresshoff, P. M., and Rolfe, B. (1973a). Transgenosis of bacterial genes from *Escherichia coli* to cultures of haploid *Lycopersicon esculentum* and

Arabidopsis thaliana plant cells. In *The Biochemistry of Gene Expression in Higher Organisms,* ed. J. Pollak and J. W. Lee, pp. 21–37. Australia and New Zealand Book Co, Artarmon.

Doy, C. H., Gresshoff, P. M., and Rolfe, B. (1973b). Biological and molecular evidence for the transgenosis of genes from bacteria to plant cells. *Proc. Natl. Acad. Sci. U.S.A. 70:*723–6.

Drummond, M. H., Gordon, M. P., Nester, E. W., and Chilton, M. D. (1977). Foreign DNA of bacterial plasmid origin is transcribed in crown gall tumors. *Nature (London) 269:*535–6.

Dudits, D., Hadlaczky, G., Bajszár, G. Y., Koncz, C., Lázár, G., and Horváth, G. (1979). Plant regeneration from intergeneric cell hybrids. *Plant Sci. Lett. 15:*101–12.

Dudits, D., Hadlaczky, G., Lévi, E., Fejer, O., Haydu, Z., and Lázár, G. (1977). Somatic hybridization of *Daucus carota* and *D. capillifolius* by protoplast fusion. *Theor. Appl. Genet. 51:*127–32.

Dunham, V. L., and Bryan, J. K. (1969). Synergistic effects of metabolically related amino acids on the growth of a multicellular plant. *Plant Physiol. 44:*1601–8.

Dunham, V. L., and Bryan, J. K. (1971). Synergistic effects of metabolically related amino acids on the growth of a multicellular plant. 2. Studies of ^{14}C-amino acid incorporation. *Plant Physiol. 47:*91–7.

Dyson, W. H., and Hall, R. H. (1972). N^6-(Δ^2-Isopentenyl)adenosine: its occurrence as a free nucleoside in an autonomous strain of tobacco tissue. *Plant Physiol. 50:*616–21.

Eichenberger, M. E. (1951). Sur une mutation survenue dans une culture de tissus de carotte. *C. R. Soc. Biol. 145:*239–40.

Engvild, K. C., Linde-Laursen, I., and Lundqvist, A. (1972). Anther culture of *Datura innoxia:* flower bud stage and embryoid level of ploidy. *Hereditas 72:*331–2.

Eriksson, T. (1967). Effects of ultraviolet and X-ray radiation on *in vitro* cultivated cells of *Haplopappus gracilis. Physiol. Plant. 20:*507–18.

Evans, D. A., and Gamborg, O. L. (1979). Effects of para-fluorophenylalanine on ploidy levels of cell suspension cultures of *Datura innoxia. Environ. Exp. Bot. 19:*269–75.

Evans, D. A., Wetter, L. R., and Gamborg, O. L. (1980). Somatic hybrid plants of *Nicotiana glauca* with *Nicotiana tabacum* obtained by protoplast fusion. *Physiol. Plant. 48:*225–30.

Fernandez, S. M., Lurquin, P. F., and Kado, C. I. (1978). Incorporation and maintenance of recombinant DNA plasmid vehicles pBR313 and pCR1 in plant protoplasts. *F.E.B.S. Lett. 87:*277–82.

Filner, P. (1966). Regulation of nitrate reductase in cultured tobacco cells. *Biochim. Biophys. Acta 118:*299–310.

Fincham, J. R. S., and Sastry, G. R. K. (1974). Controlling elements in maize. *Annu. Rev. Genet. 8:*15–50.

Flashman, S. M., and Filner, P. (1978). Selection of tobacco cell lines resistant to selenoamino acids. *Plant Sci. Lett. 13:*219–29.

Fox, J. E. (1963). Growth factor requirements and chromosome number in tobacco tissue cultures. *Physiol. Plant. 16:*793–803.

Gamborg, O. L. (1976). Plant protoplast isolation, culture, and fusion. In *Cell Genetics in Higher Plants,* ed. D. Dudits, G. L. Farkas, and P. Maliga, pp. 107–27. Akadémiai Kiadó, Budapest.

Gamborg, O. L. (1977). Protoplasts in genetic modification of plants. In *La Culture des tissus et des cellules des vegetaux,* ed. R. J. Gautheret, pp. 178–85. Masson, Paris.

Gathercole, R. W. E., and Street, H. E. (1976). Isolation, stability, and biochemistry of a *p*-fluorophenylalanine-resistant cell line of *Acer pseudoplatanus* L. *New Phytol. 77:*29–41.

Gathercole, R. W. E., and Street, H. E. (1978). A *p*-fluorophenylalanine-resistant cell line of sycamore with increased contents of phenylalanine, tyrosine, and phenolics. *Z. Pflanzenphysiol. 89:*283–7.

Gautheret, R. J. (1946). Comparaison entre l'action de l'acide indole-acétique et celle du *Phytomonas tumefaciens* sur la croissance des tissu végétaux. *C. R. Soc. Biol. 140:*169–71.

Gautheret, R. J. (1950). Remarques sur les besoins nutritifs des cultures de tissus de *Salix caprea. C. R. Soc. Biol. 144:*173–4.

Gautheret, R. J. (1955). The nutrition of plant tissue cultures. *Annu. Rev. Plant Physiol. 6:*433–84.

Gengenbach, B. G., and Green, C. E. (1975). Selection of T-cytoplasm maize callus cultures resistant to *Helminthosporium maydis* race T pathotoxin. *Crop Sci. 15:*645–9.

Gengenbach, B. G., Green, C. E., and Donovan, C. M. (1977). Inheritance of selected pathotoxin-resistance in maize plants regenerated from cell cultures. *Proc. Natl. Acad. Sci. U.S.A. 74:*5113–7.

Gengenbach, B. G., Walter, T. J., Green, C. E., and Hibberd, K. A. (1978). Feedback regulation of lysine, threonine, and methionine biosynthetic enzymes in corn. *Crop Sci. 18:*472–6.

Gleba, Y. Y. (1978). Extranuclear inheritance investigated by somatic hybridization. In *Frontiers in Plant Tissue Culture 1978,* ed. T. A. Thorpe, pp. 95–102. International Association for Plant Tissue Culture, Calgary.

Gleba, Y. Y., Butenko, R. G., and Sytnik, K. M. (1975). Fusion of protoplasts and parasexual hybridization in *Nicotiana tabacum* L. (translation). *Dokl. Akad. Nauk S.S.S.R. 221:*117–19.

Gleba, Y. Y., and Hoffmann, F. (1978). Hybrid cell lines *Arabidopsis thaliana* + *Brassica campestris:* no evidence for specific chromosome elimination. *Mol. Gen. Genet. 165:*257–64.

Gleba, Y. Y., Piven, N.M., Komarnitskii, I. K., and Sytnik, K. M. (1978). Parasexual cytoplasmic *Nicotiana tabacum* + *N. debneyi* hybrids (cybrids) obtained by protoplast fusion (translation). *Dokl. Akad. Nauk S.S.S.R. 240:*225–7.

Glimelius, K., Eriksson, T., Grafe, R., and Müller, A. J. (1978). Somatic hybridization of nitrate reductase-deficient mutants of *Nicotiana tabacum* by protoplast fusion. *Physiol. Plant. 44:*273–7.

Goldberg, R. B., Hoschek, G., Kamalay, C., and Timberlake, W. E. (1978). Sequence complexity of nuclear and polysomal RNA in leaves of the tobacco plant. *Cell 14:*123–31.

Goss, S. J., and Harris, H. (1975). New method for mapping genes in human chromosomes. *Nature (London) 255:*680–4.

Green, C. E., and Phillips, R. L. (1974). Potential selection system for mutants with increased lysine, threonine, and methionine in cereal crops. *Crop Sci. 14:*827–30.

Gresshoff, P. M. (1979). Cycloheximide-resistance in *Daucus carota* cell cultures. *Theor. Appl. Genet. 54:*141–3.

164   *Bibliography*

Guha-Mukherjee, S. (1973). Genotypic differences in the *in vitro* formation of embryoids from rice pollen. *J. Exp. Bot. 24:*139–44.

Gupta, N., and Carlson, P. S. (1972). Preferential growth of haploid cells *in vitro*. *Nature (London), New Biol. 239:*86.

Gurdon, J. B. (1960). The developmental capacity of nuclei taken from differentiating endoderm cells of *Xenopus laevis*. *J. Embryol. Exp. Morphol. 8:*505–26.

Haldane, J. B. S. (1932). The time of action of genes, and its bearing on some evolutionary problems. *Am. Nat. 66:*5–24.

Haldane, J. B. S. (1941). *New Paths in Genetics*. George Allen & Unwin, London.

Han, H., Tze-ying, H., Chun-chin, T., Tsun-wen, O., and Chien-kang, C. (1978). Application of anther culture to crop plants. In *Frontiers of Plant Tissue Culture 1978,* ed. T. A. Thorpe, pp. 123–30. International Association for Plant Tissue Culture, Calgary.

Haskell, G. (1954). Pleiocotyly and differentiation within angiosperms. *Phytomorphology 4:*140–52.

Hayes, W. (1970). *The Genetics of Bacteria and Their Viruses*. Blackwell Scientific Publications, Oxford.

Heimer, Y. M., and Filner, P. (1970). Regulation of the nitrate assimilation pathway of cultured tobacco cells. 2. Properties of a variant cell line. *Biochim. Biophys. Acta 215:*152–65.

Heinz, D. J. (1973). Sugarcane improvement through induced mutations using vegetative propagules and cell culture techniques. In *Induced Mutations in Vegetatively Propagated Plants,* pp. 53–9. International Atomic Energy Agency, Vienna.

Heinz, D. J., Krishnamurthi, M., Nickell, L. G., and Maretzki, A. (1977). Cell, tissue, and organ culture in sugarcane improvement. In *Applied and Fundamental Aspects of Plant Cell, Tissue, and Organ Culture,* ed. J. Reinert and Y. P. S. Bajaj, pp. 3–17. Springer-Verlag, Berlin.

Heinz, D. J., and Mee, G. W. P. (1969). Plant differentiation from callus tissue of *Saccharum* species. *Crop Sci. 9:*346–8.

Heinz, D. J., and Mee, G. W. P. (1971). Morphologic, cytogenetic, and enzymatic variation in *Saccharum* species hybrid clones derived from callus tissue. *Am. J. Bot. 58:*257–62.

Helgeson, J. P., Kemp, J. D., Haberlach, G. T., and Maxwell, D. P. (1972). A tissue culture system for studying disease resistance: the black shank disease in tobacco callus cultures. *Phytopathology 62:*1439–43.

Henke, R. R., Wilson, K. G., McClure, J. W., and Treick, R. W. (1974). Lysine-methionine-threonine interactions in growth and development of *Mimulus cardinalis* seedlings. *Planta 116:*333–45.

Henry, S. A., Donohue, T. F., and Culbertson, M. R. (1975). Selection of spontaneous mutants by inositol starvation in yeast. *Mol. Gen. Genet. 143:*5–11.

Hess, D., Leipoldt, G., and Illg, R. D. (1979). Investigations on the lactose induction of β-galactosidase activity in callus tissue cultures of *Nemesia strumosa* and *Petunia hybrida*. *Z. Pflanzenphysiol. 94:*45–53.

Hewitt, E. J., Hucklesby, D. P., and Notton, B. A. (1976). Nitrate metabolism. In *Plant Biochemistry,* ed. J. Bonner and J. E. Varner, pp. 633–81. Academic Press, New York.

Heyn, R. F., and Schilperoort, R. A. (1973). The use of protoplasts to follow the fate of *Agrobacterium tumefaciens* DNA on incubation with tobacco cells. *Colloq. Int. C. N. R. S. 212:*385–95.

Hibberd, K. A., Walter, T. J., Green, C. E., and Gengenbach, B. G. (1980). Selection and characterization of a feedback-insensitive tissue culture of maize. *Planta 148:*183–7.

Hinnen, A., Hicks, J. B., and Fink, G. R. (1978). Transformation of yeast. *Proc. Natl. Acad. Sci. U.S.A. 75:*1929–33.

Horsch, R. B., and Jones, G. E. (1978). 8-Azaguanine-resistant variants of cultured cells of *Haplopappus gracilis. Can. J. Bot. 56:*2660–5.

Hough, B. R., Smith, M. J., Britten, R. J., and Davidson, E. H. (1975). Sequence complexity of heterogeneous nuclear RNA in sea urchin embryos. *Cell 5:*291–9.

Hsu, T. C., and Somers, C. E. (1962). Properties of L cells resistant to 5-bromodeoxyuridine. *Exp. Cell Res. 26:*404–10.

Huber, S. C., Hall, T. C., and Edwards, G. E. (1976). Differential localization of fraction I protein between chloroplast types. *Plant Physiol. 57:*730–3.

Hughes, B. G., White, F. G., and Smith, M. A. (1977). Fate of bacterial plasmid DNA during uptake by barley protoplasts. *F.E.B.S. Lett. 79:*80–4.

Hughes, B. G., White, F. G., and Smith, M. A. (1978). Fate of bacterial DNA during uptake by barley and tobacco protoplasts. *Z. Pflanzenphysiol. 87:*1–23.

Hughes, B. G., White, F. G., and Smith, M. A. (1979). Fate of bacterial plasmid DNA during uptake by barley and tobacco protoplasts. 2. Protection by poly-L-ornithine. *Plant Sci. Lett. 14:*303–10.

Ikeda, M., Ojima, K., and Ohira, K. (1979). Habituation in suspension-cultured soybean cells to thiamine and its precursors. *Plant Cell Physiol. 20:*733–9.

Johnson, C. B., Grierson, D., and Smith, H. (1973). Expression of λplac5 DNA in cultured cells of a higher plant. *Nature (London), New Biol. 244:*105–7.

Jones, G. E., and Hann, J. (1979). *Haplopappus gracilis* cell strains resistant to pyrimidine analogues. *Theor. Appl. Genet. 54:*81–7.

Kado, C. I. (1979). Host-vector systems for genetic engineering of higher plant cells. In *Genetic Engineering: Principles and Methods,* ed. J. K. Setlow and A. Hollaender, vol. 1, pp. 223–39. Plenum Press, New York.

Kameya, T. (1975). Induction of hybrids through somatic cell fusion with dextran sulfate and gelatin. *Jpn. J. Genet. 50:*235–46.

Kao, F. T., and Puck, T. T. (1968). Genetics of somatic mammalian cells. 7. Induction and isolation of nutritional mutants in Chinese hamster cells. *Proc. Natl. Acad. Sci. U.S.A. 60:*1275–81.

Kao, K. N. (1977). Chromosomal behavior in somatic hybrids of soybean-*Nicotiana glauca. Mol. Gen. Genet. 150:*225–30.

Kao, K. N., and Michayluk, M. R. (1974). A method for high-frequency intergeneric fusion of plant protoplasts. *Planta 115:*355–67.

Kao, K. N., and Michayluk, M. R. (1975). Nutritional requirements for growth of *Vicia hajastana* cells and protoplasts at a very low population density in liquid media. *Planta 126:*105–10.

Kao, K. N., Miller, R. A., Gamborg, O. L., and Harvey, B. L. (1970). Variations in chromosome number and structure in plant cells grown in suspension cultures. *Can. J. Genet. Cytol. 12:*297–301.

Kapitsa, O. S., Kulinich, A. V., and Vinetskii, Y. P. (1977). Spontaneous lac+ mutants of tobacco tissue culture cells and their use in transgenosis (translation). *Dokl. Akad. Nauk S.S.S.R. 235:*278–81.

Kapitsa, O. S., Zueva, L. V., Vinetskii, Y. P., Likhachev, V. T., Bukh, I. G., Kunakh, V. A., Legeida, V. S., and Malyuta, S. S. (1979). Nature of β-galactosidase in a culture of tobacco cells in connection with experiments on the transgenosis of the lac+ characteristic of *Escherichia coli* (translation). *Dokl. Akad. Nauk S.S.S.R. 245:*787–9.

Kasha, K., ed. (1974a). *Haploids in Higher Plants: Advances and Potential*. University of Guelph, Guelph.

Kasha, K. J. (1974b). Haploids from somatic cells. In *Haploids in Higher Plants: Advances and Potential*, ed. K. J. Kasha, pp. 67–87. University of Guelph, Guelph.

Kasperbauer, M. J., and Collins, G. B. (1972). Reconstitution of diploids from leaf tissue of anther-derived haploids in tobacco. *Crop Sci. 12:*98–101.

Kefford, N. P., and Caso, O. H. (1966). A potent auxin with unique chemical structure – 4-amino-3,5,6-trichloropicolinic acid. *Bot. Gaz. (Chicago) 127:* 159–63.

Keller, W. A., and Melchers, G. (1973). The effect of high pH and calcium on tobacco leaf protoplast fusion. *Z. Naturforsch. 28c:*737–41.

King, P. J., Potrykus, I., and Thomas, E. (1978). *In vitro* genetics of cereals: problems and perspectives. *Physiol. Veg. 16:*381–99.

Kleinhofs, A., and Behki, R. (1977). Prospects for plant genome modification by nonconventional methods. *Annu. Rev. Genet. 11:*79–101.

Kleinhofs, A., Warner, R. L., Muehlbauer, F. J., and Nilan, R. A. (1978). Induction and selection of specific gene mutations in *Hordeum* and *Pisum*. *Mutat. Res. 51:*29–35.

Koblitz, H., Schumann, U., Böhm, H., and Franke, J. (1975). Tissue cultures of alkaloid plants. 4. *Macleaya microcarpa* (Maxim.) Fedde. *Experientia 31:*768–9.

Krishnamurthi, M., and Tlaskal, J. (1974). Fiji disease-resistant *Saccharum officinarum* var. Pindar subclones from tissue cultures. *Proc. Int. Soc. Sugar-Cane Technol. 15:*130–7.

Krumbiegel, G. (1979). Response of haploid and diploid protoplasts from *Datura innoxia* Mill. and *Petunia hybrida* L. to treatment with X-rays and a chemical mutagen. *Environ. Exp. Bot. 19:*99–103.

Krumbiegel, G., and Schieder, O. (1979). Selection of somatic hybrids after fusion of protoplasts from *Datura innoxia* Mill. and *Atropa belladonna* L. *Planta 145:*371–5.

Kung, S. D., Gray, J. C., Wildman, S. G., and Carlson, P. S. (1975). Polypeptide composition of fraction I protein from parasexual hybrid plants in the genus *Nicotiana*. *Science 187:*353–5.

Kung, S. D., Sakano, K., and Wildman, S. G. (1974). Multiple peptide composition of the large and small subunits of *Nicotiana tabacum* fraction I protein ascertained by fingerprinting and electrofocusing. *Biochim. Biophys. Acta 365:*138–47.

Lacadena, J. R. (1974). Spontaneous and induced parthogenesis and androgenesis. In *Haploids in Higher Plants: Advances and Potential*, ed. K. Kasha, pp. 13–32. University of Guelph, Guelph.

Lawrence, W. J. C., and Price, J. R. (1940). The genetics and chemistry of flower colour variation. *Biol. Rev. 15:*35–58.

Lawyer, A. L., Berlyn, M. B., and Zelitch, I. (1980). Isolation and characterization of glycine hydroxamate-resistant cell lines of *Nicotiana tabacum*. In press.

Lawyer, A. L., and Zelitch, I. (1979). Inhibition of glycine decarboxylation and serine formation in tobacco by glycine hydroxamate and its effect on photorespiratory carbon flow. *Plant Physiol. 64:*706–11.

Lescure, A. M. (1969). Mutagenèse et sélection de cellules d'*Acer pseudoplatanus* L. cultivées *in vitro*. *Physiol. Veg. 7:*237–50.

Lescure, A. M. (1973). Selection of markers of resistance to base analogues in somatic cell cultures of *Nicotiana tabacum*. *Plant Sci. Lett. 1:*375–83.

Lescure, A. M., and Péaud-Lenoël, C. (1967). Production par traitement mutagène de lignées cellulaires d'*Acer pseudoplatanus* L. anergiées à l'auxine. *C. R. Acad. Sci. Ser. D 265:*1803–5.

Lester, H. E., and Gross, S. R. (1959). Efficient method for selection of auxotrophic mutants of *Neurospora. Science 129:*572.

Levy, B., Johnson, C. B., and McCarthy, B. J. (1976). Diversity of sequences in total and polyadenylated nuclear RNA from *Drosophila* cells. *Nucleic Acids Res. 3:*1777–89.

Lewin, B. (1974). *Gene Expression–2. Eucaryotic Chromosomes.* John Wiley, New York.

Lewis, W. H., and Wright, J. A. (1974). Altered ribonucleotide reductase activity in mammalian tissue culture cells resistant to hydroxyurea. *Biochem. Biophys. Res. Commun. 60:*926–33.

Lewis, W. H., and Wright, J. A. (1979). Isolation of hydroxyurea-resistant CHO cells with altered levels of ribonucleotide reductase. *Somatic Cell Genet. 5:*83–96.

Lhoas, P. (1961). Mitotic haploidization by treatment of *Aspergillus niger* diploids with para-fluorophenylalanine. *Nature (London) 190:*744.

Libbenga, K. R., and Torrey, J. G. (1973). Hormone-induced endoreduplication prior to mitosis in cultured pea root cortex cells. *Am. J. Bot. 60:*293–9.

Limberg, M., Cress, D., and Lark, K. G. (1979). Variants of soybean cells which can grow in suspension with maltose as a carbon-energy source. *Plant Physiol. 63:*718–21.

Limbourg, B., and Prevost, G. (1971). Utilisation de marqueurs génétiques en vue de l'étude de la recombinaison de cellules végétales en culture. *Colloq. Int. C.N.R.S. 193:*241–3.

Link, G., Coen, D. M., and Bogorad, L. (1978). Differential expression of the gene for the large subunit of ribulose bisphosphate carboxylase in maize leaf cell types. *Cell 15:*725–31.

Linsmaier, E. M., and Skoog, F. (1965). Organic growth factor requirements of tobacco tissue cultures. *Physiol. Plant. 18:*100–27.

Lörz, H., and Potrykus, I. (1978). Investigations on the transfer of isolated nuclei into plant protoplasts. *Theor. Appl. Genet. 53:*251–6.

Lubin, M. (1959). Selection of auxotrophic bacterial mutants by tritium-labeled thymidine. *Science 129:*838–9.

Luria, S. E., and Delbrück, M. (1943). Mutations of bacteria from virus sensitivity to virus resistance. *Genetics 28:*491–511.

Lurquin, P. F. (1977). Integration versus degradation of exogenous DNA in plants: an open question. *Prog. Nucleic Acid Res. Mol. Biol. 20:*161–207.

Lurquin, P. F. (1979). Entrapment of plasmid DNA by liposomes and their interactions with plant protoplasts. *Nucleic Acids Res. 6:*3773–84.

Lurquin, P. F., and Kado, C. I. (1977). *Escherichia coli* plasmid pBR313 insertion into plant protoplasts and their nuclei. *Mol. Gen. Genet. 154:*113–21.

Lutz, A. (1977). Les manifestations organogènes de colonies tissulaires d'origine unicellulaire issues d'une souche anergiée de tabac. In *La Culture des tissus et des cellules des végétaux,* ed. R. J. Gautheret, pp. 124–33. Masson, Paris.

Macdonald, K. D., and Pontecorvo, G. (1953). The genetics of *Aspergillus nidulans.* 2. "Starvation" technique. *Adv. Genet. 5:*159–70.

Maliga, P. (1976). Isolation of mutants from cultured plant cells. In *Cell Genetics in Higher Plants,* ed. D. Dudits, G. L. Farkas, and P. Maliga, pp. 59–76. Akadémiai Kiadó, Budapest.

Maliga, P. (1978). Resistance mutants and their use in genetic manipulation. In

Frontiers of Plant Tissue Culture 1978, ed. T. A. Thorpe, pp. 381–92. International Association for Plant Tissue Culture, Calgary.

Maliga, P., Kiss, Z. R., Nagy, A. H., and Lázár, G. (1978). Genetic instability in somatic hybrids of *Nicotiana tabacum* and *Nicotiana knightiana*. *Mol. Gen. Genet. 163:*145–51.

Maliga, P., Lázár, G., Joó, F., Nagy, A. H., and Menczel, L. (1977). Restoration of morphogenetic potential in *Nicotiana* by somatic hybridization. *Mol. Gen. Genet. 157:*291–6.

Maliga, P., Lázár, G., Sváb, Z., and Nagy, F. (1976). Transient cycloheximide-resistance in a tobacco cell line. *Mol. Gen. Genet. 149:*267–71.

Maliga, P., Márton, L., and Sz. Breznovits, A. (1973). 5-Bromodeoxyuridine-resistant cell lines from haploid tobacco. *Plant Sci. Lett. 1:*119–21.

Maliga, P., Sz. Breznovits, A., and Márton, L. (1973). Streptomycin-resistant plants from callus culture of haploid tobacco. *Nature (London), New Biol. 244:*29–30.

Maliga, P., Sz. Breznovits, A., Márton, L., and Joó, F. (1975). Non-Mendelian streptomycin-resistant tobacco mutant with altered chloroplasts and mitochondria. *Nature (London) 255:*401–2.

Maliga, P., Xuan, L. T., Dix, P. J., and Cseplo, A. (1980). Antibiotic resistance in *Nicotiana*. In press.

Malmberg, R. L. (1979a). Temperature-sensitive variants of *Nicotiana tabacum* isolated from somatic cell culture. *Genetics 92:*215–21.

Malmberg, R. L. (1979b). Regeneration of whole plants from callus culture of diverse genetic lines of *Pisum sativum* L. *Planta 146:*243–4.

Mandel, M., and Higa, A. (1970). Calcium-dependent bacteriophage DNA infection. *J. Mol. Biol. 53:*159–62.

Manney, T. R., and Mortimer, R. K. (1964). Allelic mapping in yeast by X-ray-induced mitotic reversion. *Science 143:*581–3.

Maretzki, A., and Thom, M. (1978). Characteristics of a galactose-adapted sugarcane cell line grown in suspension culture. *Plant Physiol. 61:*544–8.

Marks, G. E., and Sunderland, N. (1966). Variability in plant tissue cultures. *John Innes Inst. Annu. Rep. 57:*22–3.

Márton, L., and Maliga, P. (1975). Control of resistance in tobacco cells to 5-bromodeoxyuridine by a simple Mendelian factor. *Plant Sci. Lett. 5:*77–81.

Márton, L., Nagy, F., Gupta, K. C., and Maliga, P. (1978). 5-Bromodeoxyuridine-resistant tobacco cells incorporating the analogue into DNA. *Plant Sci. Lett. 12:* 333–41.

Márton, L., Wullems, G. J., Molendijk, L., and Schilperoort, R. A. (1979). *In vitro* transformation of cultured cells from *Nicotiana tabacum* by *Agrobacterium tumefaciens*. *Nature (London) 277:*129–31.

Mastrangelo, I. A., and Smith, H. H. (1977). Selection and differentiation of aminopterin-resistant cells of *Datura innoxia*. *Plant Sci. Lett. 10:*171–9.

Matern, U., Strobel, G., and Shepard, J. (1978). Reaction to phytotoxins in a potato population derived from mesophyll protoplasts. *Proc. Natl. Acad. Sci. U.S.A. 75:*4935–9.

Matthews, B. F., Gurman, A. W., and Bryan, J. K. (1975). Changes in enzyme regulation during growth of maize. 1. Progressive desensitization of homoserine dehydrogenase during seedling growth. *Plant Physiol. 55:*991–8.

Matthews, B. F., and Widholm, J. M. (1978). Regulation of lysine and threonine synthesis in carrot cell suspension cultures and whole carrot roots. *Planta 141:*315–21.

Matthews, P. S., and Vasil, I. K. (1976). The dynamics of cell proliferation in haploid and diploid tissues of *Nicotiana tabacum*. *Z. Pflanzenphysiol. 77:*222–36.

Matthysse, A. G., and Torrey, J. G. (1967). Nutritional requirements for polyploid mitoses in cultured pea root segments. *Physiol. Plant. 20:*661–72.

Meins, F., and Binns, A. N. (1977). Epigenetic variation of cultured somatic cells: evidence for gradual changes in the requirement for factors promoting cell division. *Proc. Natl. Acad. Sci. U.S.A. 74:*2928–32.

Melchers, G. (1977). Microbial techniques in somatic hybridization by fusion of protoplasts. In *International Cell Biology 1976–1977,* ed. B. R. Brinkley and K. R. Porter, pp. 207–15. Rockefeller University Press, New York.

Melchers, G., and Labib, G. (1974). Somatic hybridization of plants by fusion of protoplasts. 1. Selection of light-resistant hybrids of "haploid" light-sensitive varieties of tobacco. *Mol. Gen. Genet. 135:*277–94.

Melchers, G., and Sacristán, M. D. (1977). Somatic hybridization of plants by fusion of protoplasts. 2. The chromosome numbers of somatic hybrid plants of 4 different fusion experiments. In *La Culture des tissus et des cellules des végétaux,* ed. R. J. Gautheret, pp. 169–77. Masson, Paris.

Melchers, G., Sacristán, M. D., and Holder, A. A. (1978). Somatic hybrid plants of potato and tomato regenerated from fused protoplasts. *Carlsberg Res. Comm. 43:*203–18.

Mendel, R. R., and Müller, A. J. (1978). Reconstitution of NADH-nitrate reductase *in vitro* from nitrate reductase-deficient *Nicotiana tabacum* mutants. *Mol. Gen. Genet. 161:*77–80.

Mendel, R. R., and Müller, A. J. (1979). Nitrate reductase-deficient mutant cell lines of *Nicotiana tabacum*. Further biochemical characterization. *Mol. Gen. Genet. 177:*145–53.

Meredith, C. P. (1978). Selection and characterization of aluminum-resistant variants from tomato cell cultures. *Plant Sci. Lett. 12:*25–34.

Merril, C. R., Geiser, M. R., and Trigg, M. E. (1974). Transduction in mammalian cells. In *International Conference on Birth Defects,* ed. A. G. Motulsky & W. Lentz, pp. 81–91. Excerpta Medica, Amsterdam.

Mertz, E. T., Bates, L. S., and Nelson, O. E. (1964). Mutant gene that changes protein composition and increases lysine content of maize endosperm. *Science 145:*279–80.

Miflin, B. J., and Cave, P. R. (1972). The control of leucine, isoleucine, and valine biosynthesis in a range of higher plants. *J. Exp. Bot. 23:*511–16.

Miflin, B. J., and Lea, P. J. (1976). The pathway of nitrogen assimilation in plants. *Phytochemistry 15:*873–85.

Mitra, J., Mapes, M. D., and Steward, F. C. (1960). Growth and organized development of cultured cells. 4. The behavior of the nucleus. *Am. J. Bot. 47:*357–68.

Mok, M. C., Gabelman, W. H., and Skoog, F. (1976). Carotenoid synthesis in tissue cultures of *Daucus carota* L. *J. Am. Soc. Hortic. Sci. 101:*442–9.

Morel, G. (1947). Transformations des cultures de tissus de *Vigne* produites par l'hétéroauxine. *C. R. Soc. Biol. 141:*280–2.

Motoyoshi, F., and Oshima, N. (1975). Infection with tobacco mosaic virus of leaf mesophyll protoplasts from susceptible and resistant lines of tomato. *J. Gen. Virol. 29:*81–91.

Motoyoshi, F., and Oshima, N. (1977). Expression of genetically controlled resistance to tobacco mosaic virus infection in isolated tomato leaf mesophyll protoplasts. *J. Gen. Virol. 34:*499–506.

Muir, W. H. (1965). Influence of variation in chromosome number on differentiation in plant tissue cultures. In *Proceedings of the International Conference on*

Plant Tissue Culture, ed. P. R. White and A. R. Grove, pp. 485–92. McCutchan, Berkeley.

Muir, W. H., Hildebrandt, A. C., and Riker, A. J. (1958). The preparation, isolation, and growth in culture of single cells from higher plants. *Am. J. Bot. 45:*589–97.

Müller, A. J., and Grafe, R. (1978). Isolation and characterization of cell lines of *Nicotiana tabacum* lacking nitrate reductase. *Mol. Gen. Genet. 161:*67–76.

Murashige, T., and Nakano, R. (1967). Chromosome complement as a determinant of the morphogenetic potential of tobacco cells. *Am. J. Bot. 54:*963–79.

Murashige, T., and Skoog, F. (1962). A revised medium for rapid growth and bioassays with tobacco tissue cultures. *Physiol. Plant. 15:*473–97.

Nabors, M. W., Daniels, A., Nadolny, L., and Brown, C. (1975). Sodium chloride-tolerant lines of tobacco cells. *Plant Sci. Lett. 4:*155–9.

Nabors, M. W., Gibbs, S. E., Bernstein, C.S., and Meis, M. E. (1980). NaCl-tolerant tobacco plants from cultured cells, *Z. Pflanzenphysiol. 97:*13–17.

Naef, J., and Turian, G. (1963). Sur les carotenoides du tissu cambial de racine de carrotte cultivé *in vitro. Phytochemistry 2:*173–7.

Nakamura, A., Yamada, T., Kadotani, N., Itagaki, R., and Oka, M. (1974). Studies on the haploid method of breeding in tobacco. *SABRAO Journal 6:*107–31.

Nanney, D. L. (1958). Epigenetic control systems. *Proc. Natl. Acad. Sci. U.S.A. 44:*712–17

Negrutiu, I., Jacobs, M., and Cattoir, A. (1978). *Arabidopsis thaliana* L., espèce modèle en génétique cellulaire. *Physiol. Veg. 16:*365–79.

Nelson, O. E., Mertz, E. T., and Bates, L. S. (1965). Second mutant gene affecting the amino acid pattern of maize endosperm proteins. *Science 150:*1469–70.

Nickell, L. (1961). Nutrition de tissues végétaux: sur la perte des besoins en vitamine B₁ par des tissus végétaux cultiveés *in vitro. C. R. Acad. Sci. 253:*182–4.

Nickell, L., and Maretzki, A. (1972). Developmental and biochemical studies with cultured sugarcane cell suspensions. In *Proceedings of the Fourth International Fermentation Symposium. Fermentation Technology Today,* pp.681–8.

Niizeki, H. (1976). Haploid plant of *Oryza punctata* subsp. *Schweinfurthiana* produced by treatment with *p*-fluorophenylalanine. Annual Report of the Division of Genetics of the National Institute of Agricultural Sciences (Japan), p. 22.

Niizeki, H., and Oono, K. (1971). Rice plants obtained by anther culture. *Colloq. Int. C.N.R.S. 193:*251–7.

Nishi, A., Yoshida, A., Mori, M., and Sugano, N. (1974). Isolation of variant carrot cell lines with altered pigmentation. *Phytochemistry 13:*1653–6.

Nishi, T., and Mitsuoka, S. (1969). Occurrence of various ploidy plants from anther and ovary culture of rice plant. *Jpn. J. Genet. 44:*341–6.

Nitsch, J. P. (1972). Haploid plants from pollen. *Z. Pflanzenzücht. 67:*3–18.

Norstog, K., Wall, W. E., and Howland, G. P. (1969). Cytological characteristics of ten-year-old rye grass endosperm tissue cultures. *Bot. Gaz. (Chicago) 130:*83–6.

Ohyama, K. (1974). Properties of 5-bromodeoxyuridine-resistant lines of higher plant cells in liquid culture. *Exp. Cell Res. 89:*31–38.

Ohyama, K. (1976). A basis for bromodeoxyuridine-resistance in plant cells. *Environ. Exp. Bot. 16:*209–16.

Ohyama, K., Pelcher, L. E., and Schaeffer, A. (1978). DNA uptake by plant protoplasts and isolated nuclei: biochemical aspects. In *Frontiers of Plant*

Tissue Culture 1978, ed. T. A. Thorpe, pp. 75–84. International Association for Plant Tissue Culture, Calgary.

Olson, A. C., Evans, J. J., Frederick, D. P., and Jansen, E. F. (1969). Plant suspension culture media macromolecules—pectic substances, protein, and peroxidase. *Plant Physiol. 44:*1594–600.

Oono, K. (1975). Production of haploid plants of rice (*Oryza sativa*) by anther culture and their use in breeding. *Bull. Natl. Inst. Agric. Sci.* (*Japan*) *26D:*139–222.

Oono, K. (1978). Test tube breeding of rice by tissue culture. *Trop. Agric. Res. Series 11:*109–24.

Oostindiër-Braaksma, F. J., and Feenstra, W. J. (1973). Isolation and characterization of chlorate-resistant mutants of *Arabidopsis thaliana. Mutat. Res. 19:*175–85.

Oswald, T. H., Smith, A. E., and Phillips, D. V. (1977). Herbicide tolerance developed in cell suspension cultures of perennial white clover. *Can. J. Bot. 55:*1351–8.

Owens, L. D. (1979). Binding of ColE1-kan plasmid DNA by tobacco protoplasts. *Plant Physiol. 63:*683–6.

Palmer, J. E., and Widholm, J. (1975). Characterization of carrot and tobacco cell cultures resistant to *p*-fluorophenylalanine. *Plant Physiol. 56:*233–8.

Patau, K., and Das, N. K. (1961). The relation of DNA synthesis and mitosis in tobacco pith tissue cultured *in vitro. Chromosoma 11:*553–72.

Polacco, J. C. (1976). Nitrogen metabolism in soybean tissue culture. 1. Assimilation of urea. *Plant Physiol. 58:*350–7.

Polacco, J. C. (1979). Arsenate as a potential negative selection agent for deficiency variants in cultured plant cells. *Planta 146:*155–60.

Polacco, J. C., and Havir, E. A. (1979). Comparisons of soybean urease isolated from seed and tissue culture. *J. Biol. Chem. 254:*1707–15.

Polacco, J. C., Paz, E., and Carlson, P. S. (1974). Unpublished results cited in Chaleff, R. S., and Polacco, J. C. (1977).

Polacco, J. C., and Polacco, M. L. (1977). Inducing and selecting a valuable mutation in plant cell culture: a tobacco mutant resistant to carboxin. *Ann. N.Y. Acad. Sci. 287:*385–400.

Polacco, J. C., Sparks, R. B., and Havir, E. A. (1979). Soybean urease—potential genetic manipulation of agronomic importance. In *Genetic Engineering. Principles and Methods,* ed. J. K. Setlow and A. Hollaender, vol. 1, pp. 241–59. Plenum Press, New York.

Pontecorvo, G. (1958). *Trends in Genetic Analysis.* Columbia University Press, New York.

Pontecorvo, G., and Käfer, E. (1958). Genetic analysis based on mitotic recombination. *Adv. Genet. 9:*71–104.

Potrykus, I. (1973). Transplantation of chloroplasts into protoplasts of *Petunia. Z. Pflanzenphysiol. 70:*364–6.

Potrykus, I., and Hoffmann, F. (1973). Transplantation of nuclei into protoplasts of higher plants. *Z. Pflanzenphysiol. 69:*287–9.

Power, J. B., Berry, S. F., Frearson, E. M., and Cocking, E. C. (1977). Selection procedures for the production of inter-species somatic hybrids of *Petunia hybrida* and *Petunia parodii.* 1. Nutrient media and drug sensitivity complementation selection. *Plant Sci. Lett. 10:*1–6.

Power, J. B., Frearson, E. M., Hayward, C., and Cocking, E. C. (1975). Some consequences of the fusion and selective culture of *Petunia* and *Parthenocissus* protoplasts. *Plant Sci. Lett. 5:*197–207.

Power, J. B., Frearson, E. M., Hayward, C., George, D., Evans, P. K., Berry, S. F., and Cocking, E. C. (1976). Somatic hybridization of *Petunia hybrida* and *P. parodii*. *Nature (London) 263:*500–2.

Power, J. B., Sink, K. C., Berry, S. F., Burns, S. F., and Cocking, E. C. (1978). Somatic and sexual hybrids of *Petunia hybrida* and *Petunia parodii*. *J. Hered. 69:*373–6.

Puck, T. T., and Kao, F. (1967). Genetics of somatic mammalian cells. 5. Treatment with 5-bromodeoxyuridine and visible light for isolation of nutritionally deficient mutants. *Proc. Natl. Acad. Sci. U.S.A. 58:*1227–34.

Puig, J., and Azoulay, E. (1967). Étude génétique et biochemique des mutants résistant au ClO₃ (genes chlA, chlB, chlC). *C. R. Acad. Sci. Ser. D 264:*1916–18.

Radin, D. N., and Carlson, P. S. (1978). Herbicide-tolerant tobacco mutants selected *in situ* and recovered via regeneration from cell culture. *Genet. Res. 32:*85–9.

Raveh, D., Huberman, E., and Galun, E. (1973). *In vitro* culture of tobacco protoplasts: use of feeder techniques to support division of cells plated at low densities. *In Vitro 9:*216–22.

Reinert, J. (1958). Untersuchungen über die Morphogenese an Gewebekulturen. *Ber. Dtsch. Bot. Ges. 71:*15.

Reinert, J. (1959). Über die Kontrolle der Morphogenese und die Induktion von Adventivembryonen an Gewebekulturen aus Karotten. *Planta 53:*318–33.

Reinert, J. (1973). Aspects of organization—organogenesis and embryogenesis. In *Plant Tissue and Cell Culture,* ed. H. E. Street, pp. 338–55. University of California Press, Berkeley.

Ricciuti, F., and Ruddle, F. H. (1973). Assignment of nucleoside phosphorylase to D-14 and localization of X-linked loci in man by somatic cell genetics. *Nature (London), New Biol. 241:*180–2.

Sacristán, M. D. (1971). Karyotypic changes in callus cultures from haploid and diploid plants of *Crepis capillaris* (L.) Wallr. *Chromosoma 33:*273–83.

Sacristán, M. D., and Melchers, G. (1969). The caryological analysis of plants regenerated from tumorous and other callus cultures of tobacco. *Mol. Gen. Genet. 105:*317–33.

Sacristán M. D., and Wendt-Gallitelli, M. F. (1971). Transformation to auxin-autotrophy and its reversibility in a mutant line of *Crepis capillaris* callus culture. *Mol. Gen. Genet. 110:*355–60.

Sakano, K., and Komamine, A. (1978). Change in the proportion of two aspartokinases in carrot root tissue in response to *in vitro* culture. *Plant Physiol. 61:*115–18.

Savage, A. D., King, J., and Gamborg, O. L. (1979). Recovery of a pantothenate auxotroph from a cell suspension culture of *Datura innoxia* Mill. *Plant Sci. Lett. 16:*367–76.

Scheffer, R. P. (1976). Forces by which the pathogen attacks the host plant. 4.1. Host-specific toxins in relation to pathogenesis and disease resistance. In *Encyclopedia of Plant Physiology, New Series. vol. 4. Physiological Plant Pathology,* ed. R. Heitefuss and P. H. Williams, pp. 247–69. Springer-Verlag, Berlin.

Schieder, O. (1976). Isolation of mutants with altered pigments after irradiating haploid protoplasts from *Datura innoxia* Mill. with X-rays. *Mol. Gen. Genet. 149:*251–4.

Schieder, O. (1977). Hybridization experiments with protoplasts from chlorophyll-deficient mutants of some *Solanaceous* species. *Planta 137:*253–7.

Schieder, O. (1978). Somatic hybrids of *Datura innoxia* Mill. + *Datura discolor*

Bernh. and of *Datura innoxia* Mill. + *Datura stramonium* L. var. tatula L. 1. Selection and characterization. *Mol. Gen. Genet. 162:*113–19.

Scott, A. I., Mizukami, H., and Lee, S. L. (1979). Characterization of a 5-methyltryptophan-resistant strain of *Catharanthus roseus* cultured cells. *Phytochemistry 18:*795–8.

Sharp, J. D., Capecchi, N. E., and Capecchi, M. R. (1973). Altered enzymes in drug-resistant variants of mammalian tissue culture cells. *Proc. Natl. Acad. Sci. U.S.A. 70:*3145–9.

Shewry, P. R., and Miflin, B. J. (1977). Properties and regulation of aspartate kinase from barley seedlings (*Hordeum vulgare* L.). *Plant Physiol. 59:*69–73.

Sievert, R. C., and Hildebrandt, A. C. (1965). Variation within single cell clones of tobacco tissue cultures. *Am. J. Bot. 52:*742–50.

Singh, A., and Sherman, F. (1974). Association of methionine requirement with methyl mercury resistant mutants of yeast. *Nature (London) 247:*227–9.

Singh, R., and Axtell, J. D. (1973). High lysine mutant gene (*hl*) that improves protein quality and biological value of grain sorghum. *Crop Sci. 13:*535–9.

Skirvin, R. M. (1978). Natural and induced variation in tissue culture. *Euphytica 27:*241–66.

Slavik, N. S., and Widholm, J. M. (1978). Inhibition of deoxyribonuclease activity in the medium surrounding plant protoplasts. *Plant Physiol. 62:*272–5.

Smith, H., McKee, R. A., Attridge, T. H., and Grierson, D. (1975). Studies on the use of transducing bacteriophages as vectors for the transfer of foreign genes to higher plants. In *Genetic Manipulations with Plant Material,* ed. L. Ledoux, pp. 551–63. Plenum Press, New York.

Smith, H. H., Kao, K. N., and Combatti, N. C. (1976). Interspecific hybridization by protoplast fusion in *Nicotiana. J. Hered. 67:*123–8.

Smith, S. M., and Street, H. E. (1974). The decline of embryogenic potential as callus and suspension cultures of carrot *(Daucus carota* L.*)* are serially subcultured. *Ann. Bot. (London) 38:*223–41

Sopory, S. K., and Rogan, P. G. (1977). Induction of pollen divisions and embryoid formation in anther cultures of some dihaploid clones of *Solanum tuberosum. Z. Pflanzenphysiol. 80:*77–80.

Stacey, K. A., and Simson, E. (1965). Improved method for the isolation of thymine-requiring mutants of *Escherichia coli. J. Bacteriol. 90:*554–5.

Sternheimer, E. P. (1954). Method of culture and growth of maize endosperm *in vitro. Bull. Torrey Bot. Club 81:*111–13.

Steward, F. C., Mapes, M. O., and Mears, K. (1958). Growth and development of cultured cells. 2. Organization in cultures grown from freely suspended cells. *Am. J. Bot. 45:*705–8.

Stewart, W. W. (1971). Isolation and proof of structure of wildfire toxin. *Nature (London) 229:*174–8.

Stickland, R. G., and Sunderland, N. (1972). Production of anthocyanins, flavonols, and chlorogenic acids by cultured callus tissues of *Haplopappus gracilis. Ann. Bot. (London) 36:*443–57.

Stouthamer, A. H. (1969). A genetical and biochemical study of chlorate-resistant mutants of *Salmonella typhimurium. Antoine van Leeuwenhoek; J. Microbiol. Serol. 35:*505–21.

Straus, J. (1958). Spontaneous changes in corn endosperm tissue cultures. *Science 128:*537–8.

Street, H. E. (1968). The induction of cell division in plant suspension cultures. In *Les Cultures de tissus de plantes,* pp. 177–93. Colloques Internationaux du C.N.R.S., Strasbourg.

Street, H. E. (1977). Cell (suspension) cultures—techniques. In *Plant Tissue and Cell Culture,* ed. H. E. Street, pp. 61–102. University of California Press, Berkeley.

Strogonov, B. P. (1970). *Structure and Function of Plant Cells in Saline Habitats.* 1973 Israel Program for Scientific Translations, Halsted Press, John Wiley, New York.

Stuart, R., and Street, H. E. (1969). Studies on the growth in culture of plant cells. 4. The initiation of division in suspensions of stationary-phase cells of *Acer pseudoplatanus* L. *J. Exp. Bot. 20:*556–71.

Stuart, R., and Street, H. E. (1971). Studies on the growth in culture of plant cells. 10. Further studies on the conditioning of culture media by suspensions of *Acer pseudoplatanus* L. cells. *J. Exp. Bot. 22:*96–106.

Sugano, N., Miya, S., and Nishi, A. (1971). Carotenoid synthesis in a suspension culture of carrot cells. *Plant Cell Physiol. 12:*525–31.

Sunderland, N. (1973). Nuclear cytology. In *Plant Tissue and Cell Culture,* ed. H. E. Street, pp. 161–90. University of California Press, Berkeley.

Sunderland, N. (1977). Nuclear cytology. In *Plant Tissue and Cell Culture,* ed. H. E. Street, pp. 177–205. University of California Press, Berkeley.

Sunderland, N., Collins, G. B., and Dunwell, J. M. (1974). The role of nuclear fusion in pollen embryogenesis of *Datura innoxia* Mill. *Planta 117:*227–41.

Sunderland, N., and Dunwell, J. M. (1977). Anther and pollen culture. In *Plant Tissue and Cell Culture,* ed. H. E. Street, pp. 223–65. University of California Press, Berkeley.

Sung, Z. R. (1976). Mutagenesis of cultured plant cells. *Genetics 84:*51–7.

Sung, Z. R. (1979). Relationship of indole-3-acetic acid and tryptophan concentrations in normal and 5-methyltryptophan-resistant cell lines of wild carrots. *Planta 145:*339–45.

Suzuki, M., and Takebe, I. (1976). Uptake of single-stranded bacteriophage DNA by isolated tobacco protoplasts. *Z. Pflanzenphysiol. 78:*421–33.

Suzuki, M., and Takebe, I. (1978). Uptake of double-stranded bacteriophage DNA by isolated tobacco leaf protoplasts. *Z. Pflanzenphysiol. 89:*297–311.

Tabata, M., Ogino, T., Yoshioka, K., Yoshikawa, N., and Hiraoka, N. (1978). Selection of cell lines with higher yields of secondary products. In *Frontiers of Plant Tissue Culture 1978,* ed. T. A. Thorpe, pp. 213–22. International Association for Plant Tissue Culture, Calgary.

Tatum, E. L., Barratt, R. W., and Cutter, U. M. (1949). Chemical induction of colonial paramorphs in *Neurospora* and *Syncephalastrum. Science 109:*509–11.

Tazawa, M., and Reinert, J. (1969). Extracellular and intracellular chemical environments in relation to embryogenesis *in vitro. Protoplasma 68:*157–73.

Tokarev, B. I., and Shumnyi, V. K. (1977). Detection of barley mutants with low nitrate reductase activity after seed treatment with ethylmethane sulfonate. *Genetika (Moscow) 13:*2097–103.

Torrey, J. G. (1965). Cytological evidence of cell selection by plant tissue culture media. In *Plant Tissue Culture,* ed. P. R. White and A. R. Grove, pp. 473–83. McCutchan Publishing, Berkeley.

Torrey, J. G. (1967). Morphogenesis in relation to chromosomal constitution in long-term plant tissue cultures. *Physiol. Plant. 20:*265–75.

Tulecke, W. (1959). The pollen cultures of C. D. LaRue: a tissue from the pollen of *Taxus. Bull. Torrey Bot. Club 86:*283–9.

Tulecke, W. (1960). Arginine-requiring strains of tissue obtained from *Ginkgo* pollen. *Plant Physiol. 35:*19–24.

Tumanov, I. I., Butenko, R. G., Oglevets, I. V., and Smetyuk, V. V. (1977). Increasing the frost-resistance of a spruce callus tissue by freezing out the less resistant cells (translation). *Sov. Plant Physiol. 24:*728–32.

Turgeon, R., Wood, H. N., and Braun, A. C. (1976). Studies on the recovery of crown gall tumor cells. *Proc. Natl. Acad. Sci. U.S.A. 73:*3562–64.

Uchimiya, H., and Murashige, T. (1977). Quantitative analysis of the fate of exogenous DNA in *Nicotiana* protoplasts. *Plant Physiol. 59:*301–8.

Umiel, N. (1979). Streptomycin-resistance in tobacco. 3. A test on germinating seedlings indicates cytoplasmic inheritance in the St-R701 mutant. *Z. Pflanzenphysiol. 92:*295–301.

Umiel, N., and Goldner, R. (1976). Effects of streptomycin on diploid tobacco callus cultures and the isolation of resistant mutants. *Protoplasma 89:*83–9.

Vasil, V., and Hildebrandt, A. C. (1965). Differentiation of tobacco plants from single isolated cells in microcultures. *Science 150:*889–92.

Venketeswaran, S. (1965). Studies on the isolation of green pigmented callus tissue of tobacco and its continued maintenance in suspension cultures. *Physiol. Plant. 18:*776–89.

Waddington, C. H. (1940). *Organisers and Genes.* Cambridge University Press, Cambridge.

Wallin, A., Glimelius, K., and Eriksson, T. (1974). The induction of aggregation and fusion of *Daucus carota* protoplasts by polyethylene glycol. *Z. Pflanzenphysiol. 74:*64–80.

Wallin, A., Glimelius, K., and Eriksson, T. (1978). Enucleation of plant protoplasts by cytochalasin B. *Z. Pflanzenphysiol. 87:*333–40.

Wardlaw, C. W. (1970). *Cellular Differentiation in Plants and Other Essays.* Manchester University Press, Manchester.

Waters, L. C., and Dure, L. S. (1966). Ribonucleic acid synthesis in germinating cotton seeds. *J. Mol. Biol. 19:*1–27.

Weber, G., and Lark, K. G. (1979). An efficient plating system for rapid isolation of mutants from plant cell suspensions. *Theor. Appl. Genet. 55:*81–6.

Wenzel, G., Hoffmann, F., and Thomas, E. (1977). Increased induction and chromosome doubling of androgenetic haploid rye. *Theor. Appl. Genet. 51:* 81–6.

Wetter, L. R. (1977). Isozyme patterns in soybean-*Nicotiana* somatic hybrid cell lines. *Mol. Gen. Genet. 150:*231–5.

White, P. R. (1954). *The Cultivation of Animal and Plant Cells.* Ronald Press, New York.

Widholm, J. M. (1972a). Cultured *Nicotiana tabacum* cells with an altered anthranilate synthetase which is less sensitive to feedback inhibition. *Biochim. Biophys. Acta 261:*52–8.

Widholm, J. M. (1972b). Anthranilate synthetase from 5-methyltryptophan-susceptible and -resistant cultured *Daucus carota* cells. *Biochim. Biophys. Acta 279:*48–57.

Widholm, J. M. (1972c). Tryptophan biosynthesis in *Nicotiana tabacum* and *Daucus carota* cell cultures: site of action of inhibitory tryptophan analogs. *Biochim. Biophys. Acta 261:*44–51.

Widholm, J. M. (1974a). Selection and characteristics of biochemical mutants of cultured plant cells. In *Tissue Culture and Plant Science,* ed. H. E. Street, pp. 287–99. Academic Press, New York.

Widholm, J. M. (1974b). Cultured carrot cell mutants: 5-methyltryptophan-resistance trait carried from cell to plant and back. *Plant Sci. Lett. 3:*323–30.

Widholm, J. M. (1976). Selection and characterization of cultured carrot and tobacco cells resistant to lysine, methionine, and proline analogs. *Can. J. Bot. 54:*1523–9.

Widholm, J. M. (1977a). Relation between auxin-autotrophy and tryptophan accumulation in cultured plant cells. *Planta 134:*103–8.

Widholm, J. M. (1977b). Selection and characterization of amino acid analog resistant plant cell cultures. *Crop Sci. 17:*597–600.

Widholm, J. M. (1978a). Regeneration of plants from 5-methyltryptophan-resistant tobacco cell cultures. In *Abstracts of the Fourth International Congress of Plant Tissue and Cell Culture,* p. 138. International Association for Plant Tissue Culture, Calgary.

Widholm, J. M. (1978b). Selection and characterization of a *Daucus carota* L. cell line resistant to four amino acid analogues. *J. Exp. Bot. 29:*1111–16.

Wigler, M., Pellicer, A., Axel, R., and Silverstein, S. (1979). Transformation of mammalian cells. In *Genetic Engineering. Principles and Methods,* ed. J. K. Setlow and A. Hollaender, vol. 1, pp. 51–72. Plenum Press, New York.

Wong, K. F., and Dennis, D. T. (1973). Aspartokinase in *Lemna minor* L. Studies on the *in vivo* and *in vitro* regulation of the enzyme. *Plant Physiol. 51:*327–31.

Yamada, Y., Sato, F., and Hagimori, M. (1978). Photoautotrophism in green cultured cells. In *Frontiers of Plant Tissue Culture 1978,* ed. T. A. Thorpe, pp. 453–62. International Association for Plant Tissue Culture, Calgary.

Yeoman, M. M. (1973). Tissue (callus) cultures—techniques. In *Plant Tissue and Cell Culture,* ed. H. E. Street, pp. 31–58. University of California Press, Berkeley.

Yurina, N. P., Odintsova, M. S., and Maliga, P. (1978). An altered chloroplast ribosomal protein in a streptomycin-resistant tobacco mutant. *Theor. Appl. Genet. 52:*125–8.

Zamski, E., and Umiel, N. (1978). Streptomycin-resistance in tobacco. 2. Effects of the drug on the ultrastructure of plastids and mitochondria in callus cultures. *Z. Pflanzenphysiol. 88:*317–25.

Zelcer, A., Aviv, D., and Galun, E. (1978). Interspecific transfer of cytoplasmic male sterility by fusion between protoplasts of normal *Nicotiana sylvestris* and X-ray irradiated protoplasts of male sterile *N. tabacum. Z. Pflanzenphysiol. 90:*397–407.

Zenk, M. H. (1974). Haploids in physiological and biochemical research. In *Haploids in Higher Plants: Advances and Potential,* ed. K. Kasha, pp. 339–53. University of Guelph, Guelph.

Zenk, M. H., El-Shagi, H., Arens, H., Stöckigt, J., Weiler, E. W., and Deus, B. (1977). Cell lines of *Catharanthus* producing high levels of alkaloids. In *Plant Tissue Culture and Its Bio-technological Application,* ed. W. Barz, E. Reinhard, and M. H. Zenk, pp. 27–43. Springer-Verlag, Berlin.

Index

177